W9-AFS-221

WITHDRAWN

Universal Life

Universal Life

An Inside Look Behind the Race to Discover Life Beyond Earth

ALAN BOSS

OXFORD
UNIVERSITY PRESS

Oxford University Press is a department of the University of Oxford. It furthers the University's objective of excellence in research, scholarship, and education by publishing worldwide. Oxford is a registered trade mark of Oxford University Press in the UK and certain other countries.

Published in the United States of America by Oxford University Press
198 Madison Avenue, New York, NY 10016, United States of America.

Library of Congress Cataloging-in-Publication Data
Names: Boss, Alan, 1951– author.
Title: Universal life : an inside look behind the race to
discover life beyond earth / Alan Boss.
Description: New York, NY : Oxford University Press, [2019] | Includes index.
Identifiers: LCCN 2018012947 | ISBN 9780190864057
Subjects: LCSH: Habitable planets. | Exobiology.
Classification: LCC QB820.B688 2018 | DDC 523.2/4—dc23
LC record available at https://lccn.loc.gov/2018012947

1 3 5 7 9 8 6 4 2

Printed by Sheridan Books, Inc., United States of America

TO BILL BORUCKI

the physicist who answered Fermi's paradox with "everywhere"

CONTENTS

PROLOGUE

Up, up, and away!

Superman, circa 1952–1958

After decades of painstaking planning and preparation, NASA's first dedicated exoplanet detection mission, the Kepler Space Telescope, was launched on March 6, 2009, from pad SLC-17B at the Cape Canaveral Air Force Station. Kepler's launch on a Delta II rocket was a spectacular sight on a dark, humid Florida night, as the bright light of the Delta II's rocket engines arced first upward and then eastward to disappear over the Atlantic Ocean. The engines' distant roar was reduced to a whisper, and then to complete silence, as the Kepler Mission team, assembled in a public park south of the Cape, cheered the rocket on its ride to orbit. A few gasped as the Delta II solid fuel booster rockets fell away from the first stage main engine, thinking the entire rocket had failed, but the Delta II continued to function nominally. All three stages had fired and been jettisoned within an hour after launch. Kepler was on its way.

Kepler would not orbit around the Earth, like the Hubble Space Telescope, but around the Sun, like the Spitzer Space Telescope, in an Earth-trailing orbit that would allow Kepler to stare without interruption at a large field of stars in our Milky Way galaxy. Hubble and Spitzer had performed splendid observations of newly discovered exoplanets in spite of their not having been designed to accommodate the peculiar demands of studying exoplanets. Kepler, on the other hand, was designed from the start to determine the frequency and size of rocky, terrestrial planets by searching for the tiny dips in a star's brightness if it should have a planet with an orbit that allowed the planet to pass in front of the star and block its light. William Borucki, the mission's originator and science team leader, had conceived of the exotic possibility of what Kepler could do over 25 years earlier and had spent much of the intervening decades struggling to make his personal dream a reality.

The Kepler Mission was now underway at last, with 7 years of fuel on board, more than enough for the 3.5-year prime mission of looking for Earth-like planets among the millions of stars in the northern constellations Lyra and Cygnus (the Swan). Kepler's successful launch meant that it was only a matter of a few more years before we would know just how many Earth-like planets exist in our galaxy. A revolution in thinking about our place in the universe was about to occur, one way or the other, depending on what Kepler found. Are potentially habitable Earths commonplace or rare? Are we therefore likely to be alone in the universe, or could life be widespread? Only Kepler could decide the correct answers to these vexing questions.

ACKNOWLEDGMENTS

This book is the third in a series that began with *Looking for Earths: The Race to Find New Solar Systems* (LFE, 1998) and continued with *The Crowded Universe: The Search for Living Planets* (TCU, 2009). None of these books would have been possible without the steady support and unwavering interest of my agent, Gabriele Pantucci, and the staff at Artellus Limited in London, principally Leslie Gardner. I owe both Gabriele and Leslie my profound thanks for making this third book in the series a reality.

Alan Boss
Washington, DC
July 23, 2018

1

Don't Take No for an Answer

There is no unequivocal evidence for the existence of planets outside the Solar System.

—David C. Black, Editor, 1980, Project Orion, NASA SP-436

Bill Borucki never bothered to get a PhD degree. A master's degree in physics from the University of Wisconsin in 1962 was all he needed to get a job at the NASA Ames Research Center in Mountain View, California, working on the design and testing of the materials that would form the heat shields protecting the Apollo spacecraft and their astronauts on their return to Earth from journeys to the Moon. Well after the Apollo program ended in 1972, Borucki earned a second master's degree in 1982, this time in meteorology, from San Jose State University, a short distance south of Ames. Armed with that second credential, Borucki undertook investigations of the possibility of lightning on Solar System planets, such as Venus, looking for hints of lightning strikes in the data returned to Earth by NASA's robotic spacecraft, which were busily engaged in exploring every planet in the Solar System.

Borucki (see Figure 1.1) worked for decades in the Theoretical and Planetary Studies Branch at Ames where the union card usually consists of a PhD in a hard science. Since PhD scientists do not normally parade their doctoral degrees, and office name plates do not identify the doctoral status of the occupants, Bill's lack of the proper union card was not well known in the Branch, certainly not during my time there as a postdoctoral fellow from 1979 to 1981. The day that Bill learned that he was elected to membership in the American Academy of Arts and Sciences in 2017, I sent him a congratulatory email. Bill replied that he was afraid that he had been given this new honor under false pretenses, as he lacked a PhD, and the FedEx package sent to him with the announcement of the honor was mistakenly addressed to Dr. William J. Borucki. I hurriedly reassured him that no, he had indeed been elected to membership in the American Academy on the basis of the outstanding scientific discoveries made possible by his brainchild, the Kepler Space Telescope, PhD or no PhD.

Figure 1.1 William J. Borucki, the pioneer of space transit photometry, leader of NASA's Kepler Space Telescope, and discoverer of thousands of new worlds (Courtesy of NASA).

The Space Science Division at NASA Ames was a hotbed of activities on audacious projects in the 1970s. Workshops were held in 1975 and 1976 about launching serious searches for extraterrestrial intelligence (SETI), including the Project Orion workshop that was dedicated to imagining how to detect the presence of planets in orbit around other stars. Extrasolar planets were purely hypothetical at the time, in spite of a number of failed efforts, such as the search for a Jupiter-mass planet around Barnard's star. This putative planet had been claimed to exist in 1962 by astronomer Peter van de Kamp, whose 25 years of photographic images of Barnard's star seemed to indicate that the star was wobbling back and forth as it moved across the sky, as if the star had a Jupiter-mass companion that was forcing this periodic motion. A decade later, that astrometric wobble was shown to be a spurious artifact caused by seemingly small changes in the 24-inch telescope's refracting lens and in the photographic emulsions used to take the images. The Project Orion workshop was devoted to trying to figure out how to build a telescope that would be able to detect such miniscule astrometric wobbles induced by unseen planets, with the participants settling on a gigantic pair of telescopes, called interferometers, that should do the trick.

In this infectious Ames environment, Borucki caught the exoplanet bug. But rather than pursue the same colossal ambitions as the Project Orion participants, Bill took off in another direction altogether. A psychologist at Cornell University, Frank Rosenblatt, motivated in part by Carl Sagan, had published a paper in 1971 in the planetary science journal *Icarus* describing how extrasolar planets could be detected by measuring the tiny shift in color of a star's light if a planet passed in front of the star. Besides dimming the light from the star by a small amount, the planet would cause the star to appear slightly bluer as the planet crossed the edges of the star and slightly redder as it crossed the center, as the edges of stars are redder than their centers. Borucki quashed this idea in a 1984 paper that pointed out several difficulties with trying to measure such incredibly small color changes with the 0.48-m-size (20-inch), ground-based telescope Rosenblatt had envisioned as being adequate, the most important being the deleterious effect of the Earth's atmosphere on such delicate measurements. Rosenblatt never had a chance to respond to Borucki's criticism, as he died in a boating accident shortly before his 1971 paper was published. Rosenblatt's interests ranged from artificial intelligence to SETI, and undoubtedly he would have been proud to know that his 1971 paper helped to launch Borucki's dream.

Borucki instead proposed a 1-m-size (40-inch) *space* telescope equipped with a then state-of-the-art device called a photometer, which could measure the small changes in brightness of a star when a giant planet like Jupiter passed in front of it. A Jupiter-size planet is ten times smaller in diameter than a star like our Sun, and so would dim the Sun's light by an amount equal to the surface area of the Sun that is blotted out by Jupiter, which is about 1%. A photometer that was accurate to 0.1% should then be able to detect Jupiter-like planets as they orbited around stars like the Sun, provided that the exoplanet's orbit happened to be aligned along the direction of the observer. Assuming that every star has a Jupiter-like planet, and in order to have a good chance of catching a few systems where the planet's orbit was aligned so that the planet periodically transited the star (see Figure 1.2), Borucki estimated that about 10,000 stars would have to be monitored for transits continually. The yield would be about one planet per year. This modest proposal was the birth of Borucki's concept for what would become the Kepler Mission, a 0.95-m-aperture space telescope that would stare at over 150,000 stars continuously to search not just for Jupiter-size exoplanets but for *exoEarths*.

Earth is ten times smaller in diameter than a giant planet like Jupiter, and so it can only block 1% of the star's light that a Jupiter would block, or 0.01%. The state of the art for photometers in 1984 was woefully inadequate to detect such a slight dimming of a star caused by a transiting exoEarth, even if the photometer was mounted on a space telescope well above the Earth's atmosphere. The technical advance that

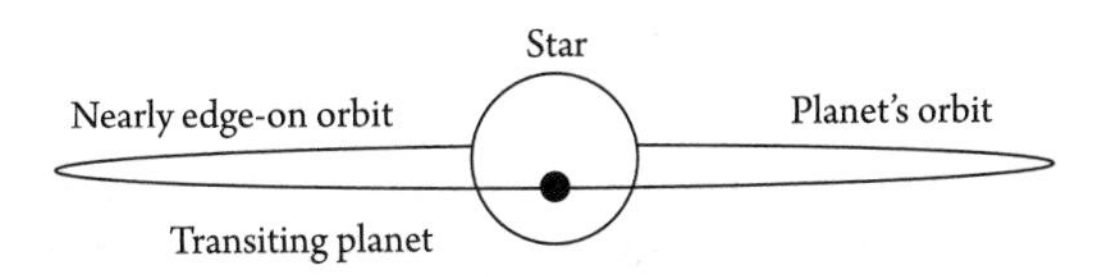

Figure 1.2 Transits occur when a planet's orbit is aligned so that the planet passes in front of its star when viewed from a telescope (e.g., on Earth), blocking and dimming the star's light.

was needed to make transit photometry capable of detecting exoEarths had been invented at Bell Labs in 1969: the charge-coupled device, or CCD, a strip of silicon coated with layers of silicon dioxide that turn the device into a semiconductor capable of efficiently converting light (photons) into electrical signals (electrons). The 2009 Nobel Prize in Physics was awarded for this invention; but by 1982, the CCD had been developed enough to begin finding its way into consumer electronics. CCDs are the basis of the digital cameras and video camcorders that first appeared in the 1980s, and regrettably led to the solipsistic phenomenon of the "selfie," once CCDs became standard features of cell phones and smart phones in the 2010s.

Borucki realized that the commercial development of the CCD offered exactly the technology he needed to make transit photometry a practical means for detecting unseen exoplanets.

The Birth of FRESIP: Working with his Ames colleague David Koch, Borucki concocted a mission concept in 1992 entitled Frequency of Earth-Sized Inner Planets, or FRESIP for short, based on a space telescope with a large array of CCD detectors. Borucki and Koch labored to convince their colleagues at Ames and elsewhere that FRESIP should be seriously considered for a future space mission, but their arguments fell on mostly deaf ears. NASA Headquarters (HQ) had chartered its first committee in 1988, the Planetary Systems Science Working Group (PSSWG), to propose a plan for the search for extrasolar planets. By 1992, we on the PSSWG had managed to draft and release the Towards Other Planetary Systems (TOPS) report after years of haggling and arguing over the various competing approaches for discovering exoplanets. The TOPS final report recommended that NASA pursue a phased program based largely on ground-based searches using van de Kamp's astrometric wobble technique, leading eventually to an astrometric space telescope capable of detecting exoplanets. FRESIP did not get even a mention in the TOPS report, though the concept of detection by Borucki's transit photometry did at least merit a few pages noting the merits and technological challenges of the approach.

The TOPS report itself fell on fairly deaf ears at NASA HQ, as there was little desire to spend substantial sums on planning and developing an expensive (by definition) *space* telescope to search for other planetary systems at a time when the only evidence we had for extrasolar planets were two burned-out cinders of orbiting rock that appeared to be making the neutron star pulsar PSR B1257+12 emit radio waves that varied in frequency with periods of exactly 25.26 and 66.54 days. The pulsar planets were interesting as oddball planetary mass objects but not good places to go if one was looking for worlds that might support life. FRESIP was designed to start that search in earnest.

Borucki and Koch got a new chance at developing FRESIP when the Discovery Program was initiated in 1992 by NASA HQ. The Discovery Program was intended to support relatively low-cost missions to Solar System objects, with teams led by a Principal Investigator (PI) from a university or government laboratory. The rankings would be made by a peer review panel of scientists and engineers rather than by fiat of agency managers. The first full competition, with 28 proposals submitted, was held in 1994. The FRESIP proposal was not accepted, largely because it was not clear if a CCD was able to measure the brightness of a star to better than the 0.01% that was needed to detect an Earth-size planet passing in front of a sun-like star. Borucki needed to work more on showing that a CCD could do the job.

The Birth of Kepler: Borucki was persuaded by Carl Sagan, SETI pioneer Jill Tarter, and David Koch that the awkward acronym FRESIP was part of the problem with his sales pitch to the Discovery Program review panels. In 1996, he accordingly renamed FRESIP as the Kepler Mission, a considerably more attractive name. Johannes Kepler (1571–1630) was a German astronomer who showed that the meticulous observations of the locations of the Solar System's planets made by Danish astronomer Tycho Brahe (1546–1601) could be explained by three basic laws of planetary motions, which have become known as Kepler's Laws. Kepler built his three laws on the prior work of Polish astronomer Nicolaus Copernicus (1473–1543), who had first made the scientific breakthrough of asserting that the planets orbit around the sun and not the other way around. Borucki might just as well have gone with honoring Copernicus instead of Kepler by renaming FRESIP as Copernicus, but an ultraviolet space telescope launched by NASA in 1972 (otherwise known by the prosaic acronym OAO-3) had already been renamed Copernicus. Tycho Brahe had a brass nose as a result of a sword duel in college that would have made for some bad puns about the nose cone on FRESIP's launch rocket. Danish engineers did name a sub-orbital human rocket mission program after Tycho Brahe, but a series of launch failures and disasters in 2010 and 2011 resulted in the Brahe program being canceled. Wisely, Borucki went with Kepler instead.

Kepler's Laws are integral to what Borucki was trying to accomplish. Kepler's Third Law is the one most often quoted in the context of exoplanetary systems, namely, that the square (second power) of the period of a planet's orbit around its star is proportional to the cube (third power) of the semi-major axis of the orbit, meaning the distance of the planet from the star. This Law says that once one knows the orbital period of a planet, one can determine the orbital distance of the planet from the star, provided that one knows the mass of the star.

This basic fact was central to the goal of the Kepler Mission: to determine the frequency of planets with both the same size (diameter) as the Earth and the same orbital distance from the central star as the Earth is from the Sun (93 million miles, or 150 million km). The Kepler Space Telescope would determine the period of a planet's orbit by timing the spacing between the periodic transits of the planet, once every orbital period. With this orbital period, and Kepler's Third Law, Borucki could then calculate how far the planet orbited from its star, and hence how much the planet would be heated by the amount of the star's light absorbed by its atmosphere and surface. This latter evaluation was central to determining whether the planet had the right temperature to allow *liquid water* to exist on the planet's surface. A hot planet would only have a steam atmosphere, with no surface water, while a cold planet would be covered with ice, assuming that abundant water was present in the first place. The presence of liquid water is the most basic and most robust requirement for a planet that is habitable and capable of evolving and supporting life. Many other constraints can be invoked as well to define a habitable world, but liquid water remains as the sine qua non.

Borucki and Koch submitted the revamped Kepler proposal to the Discovery Program again in 1996, only to be shot down once more. The same outcome resulted from the 1998 Discovery Program competition. Borucki was undaunted and prepared Kepler for the next opportunity, the 2000 Discovery Program competition.

Evidently Borucki was not the sort of person who is easily discouraged or who is willing to take no for an answer. Perhaps he was aware of the governing philosophy of the George W. Wetherill School of Management. Wetherill was a distinguished geochemist and planetary scientist who was the Director of my Carnegie Department of Terrestrial Magnetism (DTM) from 1975 to 1991. Along with Victor Safronov, Wetherill was the pioneer of theoretical modeling of how planets like Earth might form, long before we knew if there were any other planets at all beyond the Solar System. He treated being DTM Director as a part-time job, as his real love was in running the first Monte Carlo models of the collisional accumulation process of the formation of rocky planets by mutual impacts of lunar-size planetesimals. When a DTM staff member would approach Wetherill with a request, Wetherill would hesitate in looking up from his computer printouts before deciding if the request was serious enough to require his full attention. Such requests usually involved asking for money for something or other. Wetherill routinely would say, no, no, the department cannot afford that, turn around, and return to his calculations. The staff member would skulk away. Wetherill knew that if the request really was a necessary one for a good purpose, the staff member would rethink the arguments presented, return, and ask again. The second time, Wetherill might actually agree to spending the funds if a strong argument had been made. Borucki knew that he had to be at least as persistent as the staff members at DTM, especially since he was asking NASA for multiple hundreds of millions of dollars, an amount similar to the value of the entire endowment of the Carnegie Institution at the time.

Borucki had already been turned down by the Discovery Program multiple times, but in the 2000 competition, the result was finally different. Kepler was one of three proposals that were selected for further study out of the 26 submissions considered. Borucki and his team, which I had been asked to join for this round, spent the first half of 2001 performing what NASA refers to as a Phase A study, intended to more fully develop the basic concept and the technology that would be needed for the mission without actually starting to build the flight hardware (Phase C) or spending serious money. Borucki had $450,000 from NASA HQ to spend during Phase A, planning for a mission with a $299 million cost cap.

Go for Launch: On December 21, 2001, NASA HQ announced that the Kepler Mission had been selected for flight, with the launch scheduled for 2006. Borucki had triumphed at last. Now all he had to do was complete the preliminary design and technology development (Phase B); the final design and fabrication of the flight hardware (Phase C); the assembly, testing, and launch of the telescope and its spacecraft (Phase D); and then operate Kepler for a nominal 4 years of operations (Phase E). All of this was to be done for $299 million, with the launch expected in just 5 years. Clearly, the real work of the Kepler Mission was yet to be done.

Oddly enough, even though the Discovery Program was intended for planetary science investigations, the Planetary Science Division at NASA was willing to pay for a mission that studied planets but not the ones in our Solar System. At that time, it was not clear which NASA Division should have dominion over exoplanets, or even which division of the International Astronomical Union. Exoplanets were planets all right, but in order to understand them well, one needed to know a lot about their stellar hosts, and stellar astronomy is in the bailiwick of NASA's Astrophysics Division. These differences in ownership may not seem important to the uninitiated, but when a NASA Division is asked to spend $299 million, it makes a difference as to what else that money might be spent on in that Division, as each Division is expected to balance its own books without looking for funds elsewhere.

By the time 2006 arrived, Kepler was not yet launched and had suffered delays associated with difficulties in fabricating the 1.4-m primary mirror, which allowed for a wide field of view, given that the aperture was only 95 cm (38 inches), a configuration described as a Schmidt telescope design. Worse yet, the mission management had been transferred from Ames to the Jet Propulsion Laboratory (JPL) in Pasadena, California, and the resulting replan of the mission had raised the total cost to $568 million, a tad above the drop-dead cost cap of $299 million. Remarkably, NASA HQ accepted the new plan: by 2006, it was clear that exoplanets were so abundant and scientifically intriguing that Borucki and the Kepler team seemed to have a blank check from NASA HQ. The launch date slipped to November 2008.

This stable situation persisted until June 6, 2007, when Borucki sent an email to the Kepler team members letting us know that a meeting at NASA HQ a few days earlier had not gone well, not well at all. By now the cost overruns on the mission had exceeded even the $568 million figure, and Kepler needed another $42 million to finish up. This time the answer was "no." In fact, it was worse than "no": if the Kepler team did not return to NASA HQ with a new plan for moving forward that would not require HQ to pony up the funds, Kepler would now face a Cancellation Review, the formal means for shutting down a NASA flight mission.

One month later, on July 7, 2007, Borucki sent another email to the team saying that the second replan presented to NASA HQ the day before no longer requested an extra $42 million from HQ. Instead, the contractors and NASA centers involved swallowed hard and decided that they would have to eat all of the extra costs they had incurred. Kepler would live, but the nominal 4-year prime mission was now cut to only 3.5 years, barely enough time to see the required three transits of an Earth-size planet orbiting once an Earth-year in the habitable zone of a sun-like star. Three transits are the bare minimum needed to be sure that the faint dimming caused by a transit is a real, reproducible, and periodic signal.

Worse yet, Bill Borucki was demoted as the PI of the mission, replaced by a more experienced flight mission director from JPL. Borucki was now the science PI, even though those of us on the team continued to consider Bill to be the real PI.

By December 2008, the Kepler telescope and spacecraft had been fully assembled, tested as a system, and shipped to Cape Canaveral, Florida, to sit in cold storage, waiting for launch. Several more launch delays resulted from problems with the reaction wheels used to stabilize the pointing of the telescope and with the third stage of the Delta II rocket that would provide the ride to space. Kepler had the same reaction wheels that had failed in flight recently on other NASA space telescopes, so the prudent decision was made to rebuild the flight wheels in the hopes that this action would avoid the problem. Little did we know at the time how important those reaction wheels would become in a few years.

On March 6, 2009, Kepler was successfully launched into Earth-trailing orbit, with a total cost close to $640 million and a launch date that had slipped 3 years. The Kepler Space Telescope had now become the world's largest digital camera in space, with 95 megapixels, ready to perform the magic trick that Bill Borucki had dreamed of decades before. In 3.5 years, we would know just how frequent Earth-like planets are in our universe. We just had to be patient and wait a bit longer.

2

Waiting for Kepler to Deliver the Goods

A detection rate of one planet per year of observations appears possible.
—William J. Borucki and Audrey L. Summers,
1984, *Icarus*, 58, 121

Kepler's sole goal was to determine the frequency of Earth-like planets, but the mission was capable of also revolutionizing the study of stars themselves, as well as the planets in orbit about them, by studying stellar pulsations and vibrations. Kepler would pioneer a field that became known as "asteroseismology": stellar seismology would prove to be just as valuable for understanding the interiors of stars of all types and flavors as earthquake seismology is for revealing the layers and structures deep within the Earth.

Kepler was intended to be the first space telescope capable of detecting the existence of true Earth analogues, but a competing European space telescope, CoRoT (Convection, Rotation, Transits), launched several years ahead of Kepler in late 2005, was also trying to win this Nobel-quality prize. CoRoT's primary mission was stellar seismology, but the team realized early on that CoRoT could equally well search for exoplanets by Kepler's transit photometry technique, and that added scientific bonus could only enhance the chances for CoRoT's survival and success, such as during the 1999 upheaval and erosion of French support for the mission, as described in TCU. CoRoT survived that near-death experience, just as Kepler had survived one in 2007, and because of the 2-year head start, the CoRoT team seemed to have the upper hand in detecting the first Earth-like planets.

By mid-2008, the CoRoT team had discovered four hot Jupiters and a more massive, hot brown dwarf, largely gaseous bodies with such short orbital periods that they must orbit so close to their host stars as to be in danger of evaporating away. Gas giant planets like our Jupiter are physically quite similar to brown dwarfs, with the main difference being that brown dwarfs have masses more than about 13.5 times that of Jupiter. This means they are massive enough, and hence hot enough at their centers, to initiate and sustain fusion reactions involving the deuterium isotope of hydrogen early in their lifetimes, producing thermonuclear energy that allows brown dwarfs to shine brighter than lowly gas giant planets for millions of years. Hot Jupiters are fine, but the real challenge for CoRoT was to find Earth-sized bodies, 10 times smaller in size and 100 times harder to detect. By mid-2008, CoRoT had also found hints of a transiting body with a radius only about 70% larger than that of the Earth, indicative of a new class of exoplanets called *super-Earths*. If the existence of that body could be confirmed by another method, and its mass measured, then CoRoT would have accomplished the first detection of a hot super-Earth from space by transit photometry. CoRoT might score a major victory before Kepler even made it down to Cape Canaveral.

Kepler had effectively become NASA's only Terrestrial Planet Finder (TPF) space telescope (discussed in detail in TCU) in spite of the 2004 Vision for Space Exploration (VSE) advanced by President George W. Bush, which charged NASA with building a direct-imaging space telescope: that is, a TPF. Scientists and engineers at the JPL had already spent hundreds of millions of dollars developing much of the technology needed for such an ambitious space telescope and had evolved two different designs for TPF: one an optical coronagraph, which would block the glare from the planet's host star ("TPF-C"), and the other an infrared interferometer, which would use the principle of light wave interference to remove the host star's light ("TPF-I"). While TPF-C's use of visible light meant that it needed to have a roughly 8-m-diameter primary mirror in order to see an exoEarth around a nearby star, the much longer infrared wavelengths envisioned for TPF-I required a proportionately larger-diameter mirror in order to resolve such a fine detail as an exoEarth, a mirror with a size of perhaps 80 m. Such a huge, monolithic, primary mirror would be impossible to launch, so TPF-I would achieve the resolution needed by flying two 1.5-m-size mirrors, mounted on both ends of a boom some 80 m in length, or free-flying at a similar distance, and using the optical principles of interferometry to combine the mirrors into a single gigantic-size telescope capable of imaging exoEarths. Given the challenges presented by performing interferometry in space on the scale envisioned by TPF-I, the working plan in 2004 was to plan to fly TPF-C first, while development work proceeded on space interferometry, allowing TPF-I to fly later on. Being able to image and study the atmospheres of exoEarths at both visible and infrared wavelengths of light would produce the best imaginable analysis of what exactly was happening on those new worlds: would we discover evidence for life, for example water-rich planets where photosynthesis produced an oxygen-rich atmosphere?

Contrary to the grand plan envisioned by the VSE in 2004, by 2009, all NASA had to show for its several decades of planning exoplanet space telescopes was Kepler, effectively "TPF-T," intended only to detect transiting planets. Other higher-priority astrophysics missions had sapped away the resources needed to achieve the VSE, which turned out to be an unfunded mandate, as we eventually learned to our dismay.

Kepler could not directly image exoplanets or study their atmospheres. Still, Kepler was better than nothing, and Kepler would perform the precursor survey to determine the frequency of Earth-like planets that the National Academy of Sciences (NAS) Decadal Survey for astronomy and astrophysics had ruled in 2001 was necessary before NASA should contemplate developing and building anything as expensive and ambitious as a TPF, be it a TPF-C or a TPF-I. Decadal Surveys provide a periodic means for the astronomical community to settle relatively peacefully on their top science priorities for the upcoming decade. The Decadal Surveys had been accepted by both Congress and NASA HQ as the ultimate arbiter of skirmishes and battles between competing mission concepts, and even between competing areas of research; if the Decadal Survey had your back, you most likely would win any shootouts.

Now I See You, Now I Still See You: While the Kepler Mission was no substitute for a TPF-C or TPF-I, space- and ground-based telescopes began in late 2008 to make the first *reproducible* claims for the direct imaging of extrasolar planets. My first hint of this breakthrough came on November 3, 2008, when a reporter, Ron Cowen, then of *Science News*, emailed me about a press release regarding an embargoed paper to be published in the November 13, 2008, issue of *Science*. Paul Kalas of UC Berkeley and his colleagues had taken multiple images of the debris disk encircling the nearby, bright star Fomalhaut using the coronagraph on the Hubble Space Telescope (HST) and found evidence for an exoplanet, more massive than Jupiter, embedded in the ring of dust. The press release declared that the HST had achieved the "first visible light snapshot of a planet circling another star." Ron wanted to know if I agreed. My reply was considerably less positive than the tone adopted in the HST press release, a response shaped in large part by numerous similar, previous claims that had proven unfounded. Two days later, the press officer for the Space Telescope Science Institute (STScI), Ray Villard, left a phone message asking if I would be willing to participate in the press conference at NASA HQ on November 13 heralding this putative discovery, an event of the type where I had often participated in the past. A former DTM colleague of mine, Sara Seager, now of MIT, had agreed to join the press conference; but I had to decline the request, as I was scheduled to give a talk on that day at the Kepler Science Team meeting out in California.

On November 5, Ron alerted me to a second paper to be published in the same issue of *Science* as the Fomalhaut paper. Using the 10-m Keck II telescope and the 8.2-m Gemini North telescope in Hawaii, the star HR 8799 was shown by Christian Marois of the Canadian Herzberg Institute of Astrophysics to have three, and later four, gas giant planets in orbit around it, at distances from their star much greater than those of the two gas giant planets (Jupiter and Saturn, which orbit at about 5 and 10 times the Earth–Sun distance, respectively) in our Solar System. With clever imaging techniques, and repeated observations over a period of several years, Marois and his team could watch the positions of these exoplanets as they orbited around their host, a bright star about 50% more massive than our Sun and only about 30 million years old. These first three HR 8799 planets appeared to have masses 10, 9, and 6 times that of Jupiter, orbiting at distances of 24, 37, and 67 times that of the Earth from the Sun, respectively. Evidently these were relatively newly formed gas giant planets, and they were accompanied by a debris disk, the stray dust and gas leftover from the planet formation process.

The HR 8799 discovery paper seemed to me to be on more solid ground, as it involved at least three objects, a situation to be expected for gas giant planets, based on our Solar System, compared to the single object in the case of Fomalhaut. Multiple objects also lowered the odds that the objects were interlopers of some sort, perhaps faint background stars pretending to be gas giants. I told Ron that this one looked like the real enchilada, or, more accurately, the real platter of enchiladas.

Barack Obama had been elected President the previous evening, and unlike any previous Presidential candidate, he had offered to boost NASA's annual budget by $2 billion. Though most of this would likely flow to the human space flight side of the house, this seemed to be good news for NASA's scientists as well, and the two discoveries about to be published in *Science* could only further strengthen the case for a future mission to image nearby Earths.

The discovery of the HR 8799 exoplanets raised questions regarding their formation, as the conventional process for making gas giant planets, core accretion, seemed highly unlikely to be able to form such distant planets. Core accretion had trouble forming even the ice giant planets (Uranus, Neptune) in our outer Solar System. Could the HR 8799 exoplanets be the best evidence to date that there must be another mechanism capable of forming gas giant planets? The competing mechanism, disk instability, seemed able to form distant worlds even at the large distances at which the HR 8799 exoplanets orbited, provided that a planet-forming disk with sufficient mass existed. These new exoplanets began to be cited as the first possible evidence that gas giants might form in more than one way, heralding a minor revolution in theoretical models of planetary system formation.

Kepler was intended to find something quite different: relatively small, Earth-like planets with orbital periods similar to that of Earth itself. But the plan of waiting 3.5 years for planets with orbital periods of about 1 year to transit and darken their stars at least 3 times meant that shorter-period planets would produce transit events even more frequently, making them much easier to detect. The signals from multiple faint dimmings of short-period planets could be combined to produce a convincing case for their reality, allowing Kepler to probe deeply into short-period exoplanet discovery space. Ground-based detections of short-period planets, both hot Jupiters and hot super-Earths, had already established the existence of such unexpected worlds, but Kepler would provide the most complete census to date of exactly what was orbiting close to Sun-like stars, inside even where Mercury orbits the Sun. What would Kepler find there? What would Kepler's discoveries mean for our theories of the formation of rocky planets by the core accretion mechanism, where terrestrial planets form by the collisions of progressively larger rocky planetesimals and planetary embryos? Most importantly, would Earth-like planet formation be a common, or a rare, event in our galaxy?

113 Days to Launch: The Kepler Mission Science Team held a meeting at NASA Ames on November 12, 2008, where Bill Borucki told us that Kepler was completely assembled, tested, and ready to head to Cape Canaveral on January 2, 2009, for a launch scheduled at 10:47 PM on the evening of March 5, 2009. The mission had been formally moved out of the Planetary Science Division's (PSD) Discovery Program to the Astrophysics Division's (APD) newly renamed Exoplanet Exploration Program (ExEP); after two decades of planning and developing space telescope concepts (detailed at length in LFE and TCU), ExEP finally had one that was designed and dedicated to finding exoplanets.

After a few months of testing in space after launch, Kepler would start serious data collection, with the first discoveries expected to be ready by October 2009. These would be short-period planets, presumably mostly hot Jupiters, but some hot super-Earths should be in the mix as well. Testing of the 42 CCD detectors that formed the heart of Kepler showed that the desired photometric precision could be achieved: Kepler would measure the brightness of a star to about 20 parts per million, enough to detect an Earth passing in front of a Sun-like star. We heard that CoRoT, on the other hand, was having some noise problems with their detectors, which might explain CoRoT's lack of announced discoveries of hot super-Earths.

There was, however, lingering concern about the four reaction wheels on the Kepler spacecraft, rapidly spinning flywheels that act as fine steering devices for attitude control of the telescope. Because of the conservation of angular momentum of the entire spacecraft (in the absence of firing the propulsion jets), spinning up one of the wheels would cause the telescope to change the direction in which it was pointing in order to preserve the overall spin angular momentum direction. The reaction wheels employed in Kepler had been flown on previous APD missions, but several had failed well before they should have. As a result, Kepler's reaction wheels had been pulled out early in 2008, sent back to the manufacturer to be rebuilt, and then reinstalled in the spacecraft. The operational readiness review scheduled for January 20, 2008, would decide if Kepler was ready to go or not. Still, one wondered how long the reaction wheels would last: Kepler needed at least 3.5 years to achieve its primary goal, though it had enough propulsion fuel on board to last 7 years. It would be painful to keep our fingers crossed for even 3.5 years.

The next day, November 13, I gave a summary of recent developments in exoplanet research, including several dubious prior claims for direct-imaging detections of gas giant planets and brown dwarfs, as well as the considerably stronger results being published that day in *Science*. Respecting the press embargo and the authors' wishes, I showed the key discovery images of Fomalhaut and the HR 8799 system of gas giant planets, with the names of the host stars carefully blotted out. Kepler was ready to head out and determine the frequency of Earth-like planets, while the images revealed in *Science* that day hinted at what we might be able to see once we could launch a TPF-C or TPF-I.

$600 Million Down, Just $1 Billion to Go: The engineers at the JPL had spent several years and hundreds of millions of dollars figuring exactly how to detect any Earth-mass planets that might be lurking on Earth-like orbits around the Sun-like stars closest to Earth, the planets that they hoped would some day be imaged by TPF-C or TPF-I. They designed a space telescope called the Space Interferometry Mission (SIM: see LFE and TCU), which could detect these unseen planets *indirectly*—not by direct imaging of the planets, as TPF-C or TPF-I would do, but by following the gravitational effect that an orbiting planet has on the host star around which it orbits. Any star with a planetary system must orbit around the center of mass of the entire rotating system, wobbling to and fro, over and over again. SIM would use the same astrometric detection technique that had been used by Peter van de Kamp in his ill-famed studies of Barnard's star but with a precision over a thousand times better, enough to detect not only van de Kamp's putative gas giant but even Earth-mass exoplanets. SIM would not only detect these exoEarths, but it would also weigh them: astrometric planet detections are unique in the way in which they determine the complete orbital parameters and masses of exoplanets.

The SIM Science Team held a meeting on November 19, 2008, at the Naval Observatory in northwest Washington, DC. The major goal of this meeting was to plot a strategy for obtaining a high priority ranking in the next NAS Decadal Survey on astronomy and astrophysics, scheduled to begin shortly. The Decadal Survey was intended to develop a consensus within the community about how federal research dollars in astronomy (primarily from the National Science Foundation [NSF] and NASA) should be spent in the 2010–2020 decade. Getting a high ranking on the "Astro 2010" report would be essential to having any chance of being approved and funded by either agency. The main question was how ambitious of a mission should be proposed by the SIM team to Astro 2010 in order to maintain a favorable ranking. The 1991 Decadal Survey had recommended an Astrometric Interferometry Mission (AIM) concept as a high priority. The 2001 Decadal Survey had assumed that AIM, now named SIM, would continue to be developed successfully and launched as planned in 2005, and then proceeded to anoint the 8-m Next Generation Space Telescope (NGST) concept as its top priority for a large mission for the first decade of the 21st century. The SIM team was painfully aware that SIM had not in fact launched in 2005, and they needed to try to make sure that SIM might fly in the second decade of the new millennium. A reasonable price tag would be needed for SIM to survive yet another Decadal Survey.

SIM had gone through many iterations following the AIM concept, beginning with the Orbiting Stellar Interferometer (OSI), a 20-m-long interferometer first proposed in 1998 by JPL's Michael Shao (see LFE); to "SIM-Classic," a downsized, 10-m-long version; and then to a bare-bones SIM astrometry concept, mischievously termed "SIM-Lite," a 6-m-long version (see TCU). The reduced sizes meant a reduction in capability, of course, but the goal was to reduce the cost as much as possible and so keep the project alive: the project's funding had been tight for several years and was slated to decrease precipitously in the following year. The SIM project was on track to have spent about $600 million by the end of FY 2010, averaging $50 million per year over its 12-year lifetime. For comparison, the cost of building the 6.5-m James Webb Space Telescope (JWST, formerly the NGST), the top priority of the 2001 Decadal Survey, was running at about $600 million per year, making SIM's spending spree seem minor in comparison; but clearly Astro 2010 would be looking for SIM to come in a lot cheaper than JWST.

An independent cost estimate was underway for what it might cost to finish developing and testing the exacting technology needed for SIM-Lite, then in Phase B. The estimate was expected to come in at about $900 million. When the cost of a launch vehicle was factored in, as well as the funds expended to date, it looked like the JPL could deliver SIM-Lite for a total cost of about $1.6 billion. Astro 2010 would only have to come up with another $1 billion or so to support a mission that would discover and determine the mass of numerous nearby Earths. Whereas a somewhat cheaper version was also considered—SIM-PH (Planet Hunter), focused solely on exoplanets—the SIM project and JPL decided to place their bet on SIM-Lite, which would perform important astrophysics besides planet hunting. A wider appeal beyond finding nearby Earth-like exoplanets was considered critical to winning a high ranking by Astro 2010. But would even that be enough to win approval by a group not composed of exoplanet aficionados?

Hubble Hits a Home Run: An inquiry from a *Nature* reporter landed in my email inbox on November 19, 2008, asking about a claim for the detection of carbon dioxide in the atmosphere of a hot Jupiter in the HD 189733 system. The star, the 189,733rd star in the catalog of 225,300 stars compiled by Harvard astronomers in work funded by the widow of astronomer Henry Draper, is a binary star whose solar-type primary is orbited by a short-period planet named HD 189733 b. Exoplanet nomenclature obeys the astronomical convention of naming planets after their host star's name, followed by the lowercase letters b, c, d, and so forth, in the order of their discovery, should more than one exoplanet be found in the system.

HD 189733 b is a transiting exoplanet, meaning that careful observations of what happens just before, during, and after the exoplanet passes either in front of the star (a transit) or behind the star (an occultation) can reveal clues about the exoplanet's atmosphere, even though the exoplanet cannot be seen directly in the glare of its star. These clues are deduced by studying the spectrum of the planet's atmosphere, that is, the amount of light either emitted or absorbed by the gases and particles in the atmosphere, across a range of wavelengths of light. The key is to take precise measurements of the combined spectra of the planet and the star just before an occultation, and then to subtract away the purely stellar spectrum obtained during the occultation, leaving behind the planet's spectrum, buried in whatever noise was associated with the measurements (see Figure 2.1).

For a transit, the trick is to measure how much stellar light is blocked at various wavelengths, which gives information about the height of the planet's atmosphere that is able to block or transmit stellar light of differing wavelengths (see Figure 2.1).

The difficulty of taking and interpreting these measurements was exemplified by the claim made in April 2007 by Italian astronomer Giovanna Tinetti that water had been detected for the first time in the atmosphere of HD 189733 b using observations from the Spitzer Space Telescope (SST). SST is a 0.85-m infrared NASA space telescope, and Tinetti was using measurements from an SST instrument that operates in the infrared wavelength range of 3.6 to 8 microns; visible light is shorter in wavelength, extending from about 0.4 to 0.7 microns. Tinetti's claim was quickly challenged by two other groups who also used SST data, and later a third, who could not reproduce the claim for water. However, in March 2008, Tinetti and JPL's Mark Swain published a *Nature* paper claiming once again the detection of water in HD 189733 b, this time using the HST's near-infrared camera. This one was not challenged: the HST evidence was compelling proof that water had been found in an exoplanet, and the detection was confirmed with further SST observations by another group. There was even a hint of methane in the HST spectrum of HD 189733 b (see TCU for more details).

The *Nature* reporter was present at a meeting in Paris on November 19–21, 2008, and had heard Tinetti's talk about her latest HST observations of HD 189733 b. Tinetti presented strong evidence that HST was able to detect the presence of carbon dioxide in the atmosphere of an exoplanet for the first time. Carbon dioxide is considered to be one of the primary biomarkers for any habitable planet, along with water, oxygen, and methane.

Earth, Venus, and Mars all have carbon dioxide in their atmospheres. The HST detection of carbon dioxide in HD 189733 b was not for a potentially habitable world, however, but for a hot Jupiter, a gas giant orbiting so close to its host star that its atmospheric temperature is thousands of degrees. Still, the fact that carbon dioxide could be detected even on this hot Jupiter meant that another step along the road to being able to detect these biomarker molecules in the atmosphere of a nearby Earth had been taken.

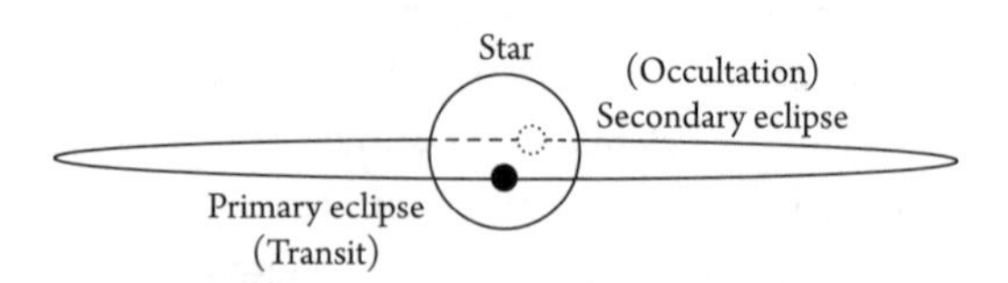

Figure 2.1 During a primary eclipse, the planet transits in front of the star, dimming the star's light, whereas during a secondary eclipse, or occultation, the planet passes behind the star so that the planet's light is blocked by the star.

The hope was that this HST discovery would help to buttress the rationale for building TPF-C and TPF-I, which could do something that HST was never designed to do: take pictures of nearby Earths. Even HST's designated successor, the JWST, could not match the exoplanet capabilities of TPF-C and TPF-I. JWST, the top priority of the 2001 Decadal Survey, was scheduled for launch in mid-2013, but its top priority was "visiting a time when galaxies were young," the subtitle of the 1997 NGST Study Team report. The NGST report had followed on the heels of the 1996 *HST and Beyond* report, which had ranked primeval galaxies as one of two major goals, along with "the detection of Earth-like planets around other stars and the search for evidence of life on them." The fact that the two major goals were numbered (1) galaxies and (2) exoEarths seemed to imply a ranking of the two, whether this ranking was intended or not. It was notable that none of the 17 august members of the *HST and Beyond* committee could be considered to be exoplanet enthusiasts, especially since the field of exoplanets did not exist in 1996 to the extent that it does now. The first confirmed exoplanet, 51 Peg b, was announced in the fall of 1995, whereas the committee had been appointed in 1993. Timing is everything, and so is the choice of committee members.

The fact that exoEarths even placed in the *HST and Beyond* race rather than winning it was considered a triumph at the time by those of us who were already used to betting on exoEarth race entries. The NGST Study Team report decreed that studying primeval galaxies would be the science goal driving the design and capabilities of the JWST, and their new concept of a segmented primary mirror meant that the JWST would not be able to image exoplanets: stray star light reflected, scattered, and diffracted off the edges of the many segmented primary mirror tiles would swamp any effort to see a faint, close-in exoplanet.

The NGST report had estimated that JWST would cost significantly less than the $2 billion cost to build HST, and that it could be built and launched by 2005. If so, there seemed to be enough money in the NASA budget in 1997 to continue with building SIM and to build and fly a Planet Finder space telescope as well, both for no more than a total of $1 billion (see LFE). No cause for alarm, right? Wrong. The SIM science team meeting in November 2008 had displayed the peril of assuming that NASA would be able to afford everything that was approved or assumed by a Decadal Survey. We could only hope that SIM would survive Astro 2010 and maybe even a TPF or two.

Europe Goes for Broke: The results of the European version of a Decadal Survey, called Astronet, were released in November 2008. Astronet called for something truly stupendous. European astronomers proposed building the European Extremely Large Telescope (E-ELT), a 42-m-diameter telescope that would dwarf the 10-m-class telescopes that were currently the world's largest ground-based telescopes. But could even affluent Europe afford the cost of the E-ELT? What would a 42-m-diameter telescope cost, even if it was assembled from a huge number of hexagonally tiled, 1.42-m mirrors, 798 to be precise, similar to the 10-m Keck telescopes on Mauna Kea? Astronet also called for a space telescope that would search for hypothetical gravitational waves: LISA, the Laser Interferometer Space Antenna. No mention was made of the European Space Agency's (ESA) previous collaborative efforts with NASA and the JPL to develop a joint version of TPF-I called the Emma Darwin x-array, named after the wife of Charles Darwin (see TCU). Exoplanets were off the European fixed-price menu, unless they could be discovered and studied somehow with the E-ELT.

The E-ELT topped the wish list of European astronomers for the upcoming decade. Their decision would put pressure on the NAS 2010 Decadal Survey, then getting underway. Would the United States continue to try to compete with Europe in dominating ground-based astronomy, or would it give up the lead, as the United States had done in the field of particle physics, canceling the Superconducting Super Collider (SSC) in 1993 and yielding to the construction of the Large Hadron Collider (LHC) on the Swiss-French border near Geneva? The United States had built the twin 8.2-m Gemini telescopes in 2000, one on Mauna Kea in Hawaii and one on Cerro Tololo in Chile, while the European Southern Observatory (ESO) had built the four equally large 8.2-m Very Large Telescopes (VLT) on Cerro Paranal in Chile, starting in 1998. Would the next U.S. Decadal Survey meet this challenge and propose astronomical parity by building something as grand as the E-ELT?

Who's in Charge Now?: With the election of President Barack Obama, NASA's Administrator, Michael Griffin, submitted his letter of resignation, as required for all political appointees, in December 2008; but he hoped that the letter would not be accepted. A number of supporters pushed for Griffin to be retained by the new Administration, but with the January 20, 2009, Presidential inauguration date fast approaching, Griffin had still not heard from Obama's transition team. Griffin had played a major role in siphoning off funds from the Science Mission Directorate to solve other problems in the Agency, and he had cut the funding for the SIM and the TPFs, apparently in order to increase the annual spending on the JWST, which was gobbling dollars at a voracious rate. While Griffin had stated in January 2008 that he too wanted to know if we are alone in the universe, he felt that NASA could not afford the SIM along with the HST, the JWST, and the rest of the APD portfolio (see TCU). Something had to give, and it would be the SIM.

Rumors began to spread that Obama would seek a new leader for NASA, perhaps a distinguished scientist. Although not a scientist, Charlie Bolden, a former NASA astronaut and Marine Corps Major General, was reported to be at the top of the list. A later rumor surfaced on January 13, 2009, that an Air Force General, who had served as one of President Obama's advisors during the campaign, had been asked to succeed Griffin; but the announcement was premature, at best.

So what if Charlie Bolden got the nod? Given the ultimate power of the NASA Administrator for whatever NASA wants to do, Bolden would be the one who controlled the future destiny of NASA's space telescopes. But what would Bolden think about finding nearby Earths, of flying the SIM and a TPF or two? No one knew for sure. Bolden had flown four times on the Space Shuttle, including the flight that placed the HST in orbit. Perhaps Bolden would have a fond memory of having helped launch arguably the most famous telescope in history and would support NASA's plans for even more capable space telescopes. We would just have to wait and see.

CoRoT Strikes First: The first email, from Jean Schneider of the Paris Observatory, arrived at 9 AM Eastern on February 3, 2009. The CoRoT team was holding an international symposium in Paris, France, and had made a major discovery. Even before Kepler could be launched, the CoRoT team announced that it had detected the smallest-radius exoplanet ever found, a world with a diameter less than twice that of the Earth.

The planet, dubbed "CoRoT-Exo-7b," lest anyone forget which telescope discovered it, was thought to have a mass somewhere in the range of 6 to 11 times that of the Earth, making it a "super-Earth." Given the uncertain mass of CoRoT-Exo-7b, however, the mean density of the planet could not be ascertained well enough to state if it was likely to be composed primarily of rock and iron, like the Earth, or if it might have significant lighter components, such as water, making it more similar to the ice giant planets, Uranus and Neptune. Since CoRoT-Exo-7b was a "hot super-Earth," however, it might then be best described as a water world rather than an ice world. The planet's atmosphere at its surface would be hotter than in any steamy sauna bath, and hence inhospitable to life, but CoRoT-Exo-7b was another step in the right direction. The CoRoT had found its first hot super-Earth, beating Kepler to the punch.

Email inquiries began to fly from reporters asking me about the latest CoRoT discovery. *Science* magazine, *Science News*, the *New York Times*, the Associated Press, *USA Today*, MSNBC, and SPACE.com all wanted their questions answered, now, please. Was this the first detection of a transiting hot super-Earth? Well, no, given that Belgian astronomer Michael Gillon and his Geneva Observatory team had announced the discovery of the first transiting hot super-Earth, Gliese 436b, in May 2007, based on ground-based observations. Gliese 436b had a mass about 23 times that of Earth, which to my mind made it a super-Earth, though with mass about half as large, CoRoT-Exo-7b was even more of a super-Earth. Was CoRoT-Exo-7b the smallest exoplanet found to date? Not necessarily, as ground-based Doppler searches, which detect exoplanets indirectly by measuring the tiny Doppler shifts in the star's spectrum as it orbits around the center of mass of the star-planet system (see Figure 2.2), had found several hot super-Earths with masses about 8 times that of Earth. These Doppler exoplanets did not transit their stars, though, so their sizes could not be measured directly, but given their masses, they might well be as small as, or smaller, than CoRoT-Exo-7b.

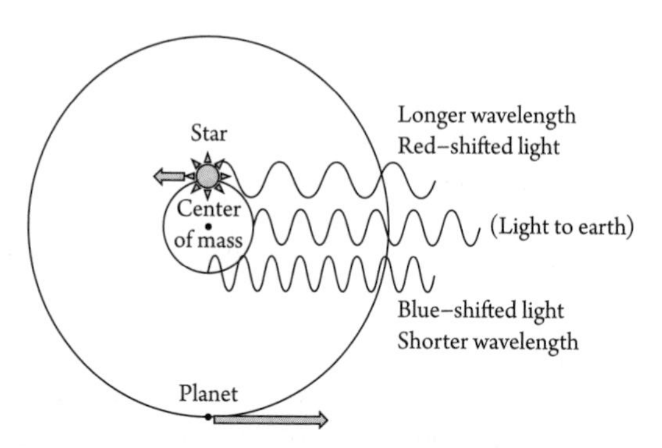

Figure 2.2 The presence of an unseen planet can be inferred from the Doppler shift of light emitted by its star, as the star orbits around the center of mass of the system.

The new CoRoT exoplanet had a direct size measurement but no mass determination, as Doppler spectroscopy had not yet been able to detect a periodic velocity wobble that could be disentangled from the spectral variations intrinsic to its host star, a billion-year-old K dwarf star. We would have to wait a bit longer to figure out just what CoRoT-Exo-7b was primarily: a rock world, a water world, or something in between.

3

Astro 2010 Comes to Bat

In preparing for battle, I have found that planning is essential, but plans are useless.

—Dwight D. Eisenhower, U.S. General and President, 1890–1969

While doing a live radio interview about TCU on February 19, 2009, an email arrived from the National Academy of Sciences inviting me to serve on the NAS Decadal Survey panel that would consider proposals for future telescopes capable of making electromagnetic observations from space, the "EOS" Program Prioritization Panel. Simply put, EOS would be charged with ranking all the ideas for space telescopes that would involve light, ranging from X-rays, through the ultraviolet and visible, to infrared wavelengths and beyond. Besides ranking the proposals, the EOS Program Prioritization Panel was charged with creating a balanced and integrated program for the field. In addition to EOS, there would be a Program Prioritization Panel dealing with space missions involving gravitational waves and particle physics, and two panels considering ground-based observations at long (radio to submillimeter) or short (visible to infrared) wavelengths. The first two panels would help to give NASA its marching orders for the 2010s, while NSF would have to respond to the priorities favored by the latter two. Although the Program Prioritization Panel inputs would be crucial to the decision process, in the end, it would be the Decadal Survey Committee itself, a group of 23 accomplished astronomers, astrophysicists, and engineers, who would make the final decisions. Roger Blandford of Stanford University would chair the Survey Committee, taking on the onerous task of deriving a consensus from the opinions of many competing voices.

The previous year I had turned down a request from the NAS to co-chair the committee overseeing U.S. research efforts on the origin and evolution of life, having had enough previous NAS committee work to last me for a long time, but this new request was different. The output of the EOS panel would have a major effect on the ranking of future NASA missions in the Decadal Survey, which meant that the EOS panel was going to be in control of the destiny of the SIM and TPF-C, the TPF concept that was judged by the JPL to be best suited for possible development and launch by 2020. I had been dreaming along with the JPL engineers and the nascent exoplanet community about missions like the SIM and the TPFs ever since 1988. How could I turn down this request to serve on the EOS panel? Not wishing to rush an important decision, I made a list of pros and cons and then slept on the NAS request. I sent my acceptance email the next day. It is often said that once you pick the members of a committee, you have effectively determined what the committee will decide to recommend. It would be interesting to see if this rule would hold true in the case of a Decadal Survey panel. Who else would be on the EOS panel?

The new Survey would be starting off with a significant amount of unfinished business, both in space and on the ground. The top ground-based priority of the 2001 Survey had been the Giant Segmented Mirror Telescope (GSMT). Two concepts had been developed for the GSMT: the Giant Magellan Telescope (GMT) and the Thirty Meter Telescope (TMT). The GMT was to be 24.5 m in diameter and located at Carnegie's Las Campanas Observatory in Chile, while the 30-m-diameter TMT would be built on Mauna Kea on the Big Island of Hawaii. Both projects were well into planning, but the NSF had not been able to support either GSMT concept to a significant extent. Clearly this 2001 Survey goal had yet to be realized, and so the GSMT would fall into the lap of the next Survey to be reconsidered. In fact, of the 20 projects listed as priorities by the 2001 Decadal Survey, only 5 had been undertaken or finished. The only large-category space mission that was underway was the JWST, with a current cost now estimated at $4.5 billion.

Over 100 proposals for space missions and ground-based telescopes had been submitted to the new Survey, ensuring that there would be considerably more grief than joy once the final selections were announced. Given the high stakes involved, literally billions of dollars, it was likely to be a treacherous competition. Considering the failure of previous Decadal Surveys to forecast the actual cost of space missions, it was also clear that the next Survey would have to do a much better job of estimating what its highest-priority missions should cost and what should be done if they blew past their cost estimates. This would not be easy.

The first meeting of the EOS panel was scheduled for early May 2009, just a few months off, with several more meetings scheduled later in 2009. The panels were to provide "an interim internal and confidential summary preliminary report of its recommended program and rankings by the fall of 2009," a little over half a year away. It was time to fasten your seat belts and get ready for the wild ride of a Decadal Survey.

An Exoplanet by Any Other Name: Bill Borucki sent out an email to the Kepler team on February 26, 2009, announcing a slight delay in Kepler's launch date. A NASA Earth Science Division (ESD) satellite launch on February 24 had ended in disaster. The payload fairing, which covers and protects the satellite during launch, had failed to detach itself; and with the unplanned extra weight of the shroud, the rocket could not make it into orbit. Instead, the $273 million satellite crashed into the ocean close to Antarctica. As a result, with an appropriate abundance of caution,

NASA was holding a review board to make sure that Kepler did not share any of the same components as the failed mission. This would result in a launch delay, but only of a single day, to the evening of March 6, 2009. Bill Borucki would have no problem waiting one more day to see his dream launched safely.

The Kepler Mission held a science team meeting in Cocoa Beach, Florida, south of Cape Canaveral, starting on Monday, March 2, a few days before the newly rescheduled launch date. We learned that the Kepler would be carried aloft by a Delta II 7925-10L rocket, the most reliable launch vehicle in the U.S. inventory, with only a few more left in stock. The first stage of the Delta II would be liquid-oxygen fueled, while nine solid-fuel rocket motors would be strapped around the bottom of the first stage. Six of these solid rocket boosters would fire during the initial lift-off phase, and then jettisoned once the Delta II was in flight and they had burned out, when the remaining three would be lighted. These pyrotechnic events once already in flight would produce moans of dread among those science team members watching from the Cape Canaveral shoreline who did not know what was happening—to them it appeared that something had gone horribly wrong, as it had with the ESD launch a month earlier.

Kepler was already on the launch pad by then, but the Delta II second stage had not yet been filled with its fuel, as the fuel was so toxic and corrosive that it would ruin the second stage if the Delta II was not launched shortly thereafter. We learned on March 2 that the Kepler Launch Readiness Review board would be holding its final meeting that evening; and if they granted their final approval, the second stage would be fueled, starting the clock on a 1-week period in which Kepler would have to launch. Their main concern was the reaction wheels that had failed prematurely on other NASA science missions; but given that Kepler's wheels had been cleaned and tested, and one had spun for over 2 billion revolutions, Kepler was likely to be declared good to go that evening.

The ESD failed mission review board had found one component in common between the Kepler launch vehicle and the doomed rocket: the ordnance that exploded and released the payload fairing. At first glance, that seemed like the worst possible component to have in common, given the failure to release the fairing, but the ordnance was a device used successfully in many prior missions and was not thought to be the reason for the ESD failure. Kepler would soon test out that review board conclusion.

At 6:35 PM that Monday evening, Bill Borucki announced to the cheers of the Science Team that Kepler had been cleared for launch at 10:49 PM (Eastern time) on Friday night, March 6. After over 20 years, Bill had just 4 more days to wait.

But First, Let's Finish the Meeting: The second day of the Science Team meeting was held on March 3, with talks by many of the 35 Science Team members about the planned follow-up work that would need to be accomplished once Kepler started finding stars that seemed to dim periodically. False positives would need to be rejected (for example, background eclipsing binary stars that could masquerade as a transiting planet around a foreground star), and planetary masses would need to be determined by Doppler spectroscopy for the transiting exoplanets Kepler would find.

After considerable debate, the Science Team took a vote on the nomenclature to be used for the exoplanets to be found by Kepler: the first one would be called Kepler-1b, the second one Kepler-2b, and so on, following the lead established by the CoRoT mission, as well as by several ground-based planet search surveys where the tradition of using the host star's astronomical name (e.g., HR 8799) to name the exoplanets (hence HR 8799 b, HR 8799 c, etc.) was supplanted by the name of the survey effort. A few days after the successful launch of Kepler, Jean Schneider, who was the first to produce and maintain an extensive catalog of extrasolar planets, sent out an email to his worldwide distribution list stating that the CoRoT Scientific Council had met and decided to change their exoplanet nomenclature from the previous system. Their hot super-Earth, initially dubbed "CoRoT-Exo-7b," would henceforth be known as "CoRoT-7b." This was the same naming rule that the Kepler science working group had voted for a few days earlier. A CoRoT science team member was present at the Cocoa Beach meeting and evidently reported back to the CoRoT Scientific Council what the Kepler team had debated and decided.

Kepler Update L+2: David Koch sent the team an email with that subject line two days after launch, noting that after a "great launch on Friday night" (see Prologue), the mission was "going smoothly." The spacecraft had crossed the orbit of the Moon and was on its way to an Earth-trailing orbit around the Sun: "There is no turning back." The operations team had contacted the spacecraft and turned on the photometer, the 95-megapixel camera, but the dust cover protecting the mirrors would not be jettisoned until L+19, so Kepler effectively still had its giant 95-cm-wide eye closed. Koch sent another team email update the next day and then a third noting that this one would be his last such email. NASA policy was that all public information about an operating space mission had to be cleared first by NASA HQ and then posted on the Kepler web site for all to read.

Kepler had not yet taken any science data, but it was already having a significant effect on the exoplanet world. A March 24 email from Gordon Walker of the University of British Columbia, pioneer of the basic Doppler technique that first enabled the spectroscopic discovery of exoplanets (see TCU), alerted me to the fact that the CoRoT 27-cm-diameter telescope had lost one of its two eyes on the day after Kepler had launched. Evidently two of CoRoT's four CCDs had failed on March 8, 2009, effectively reducing CoRoT's ability to monitor target stars for transiting planets by a factor of two. But there was no mention of this failure on the CoRoT mission's web page.

What was going on with CoRoT? Walker jokingly wondered if the Kepler team had somehow managed to sabotage CoRoT. The CoRoT–Kepler rivalry was certainly intense, but in the end, we were all scientists seeking answers to the same question: how common are Earth-like planets? We wanted to know the right answer more than we wanted to claim credit; there was fame enough to go around in the exoplanet business.

I Can See! I Can See!: A joyous email from Bill Borucki, Kepler's creator and Science Team leader, was circulated to the team late on April 7, 2009. The dust cover that protected the mirrors on Kepler and temporarily blocked its vision had been successfully opened by heating up a wire until it broke, allowing the spring-loaded dust cover to pop off and drift away from the telescope. Kepler could now take its first glimpse of the universe. Calibration of the camera would begin the next morning.

On April 16, NASA HQ released the "first light" image taken by Kepler of its target field in the constellations of Cygnus and Lyra: a stunning image indeed, roughly 10 degrees by 10 degrees in size, containing literally millions of stars. Kepler would photograph this huge region, equivalent to over 20 lunar diameters on a side, every 30 minutes for the next 3.5 years, monitoring the brightness of over 150,000 stars, searching for transiting exoEarths. In what would turn out to be a spectacular understatement, the NASA press release noted that Bill expected Kepler to find hundreds of transiting planets. Given that the mission was just getting started, it was probably better to be conservative in predicting what Kepler would find. After a few more weeks of testing, Kepler would start taking real data, science data, on May 12, and then we would see how many planets Kepler would find.

So far, Kepler was performing flawlessly, ready to begin its prime mission of counting Earths, one by one. However, ominous signs of the intentions of the new Administration were beginning to emerge. President Obama's preliminary FY 2010 budget request, released in February, had asked for an increase in NASA spending on Earth science but nothing much for astrophysics or planetary science. In fact, when the full budget request became available on May 7, we learned that the ExEP budget had been cut by almost a third, compared to outgoing President George W. Bush's FY 2009 budget request for the ExEP.

Monitoring global climate change appeared to be more important to the new President than the dramatic discoveries to be made by the ExEP or future Mars rovers. Worse yet, the rising costs of building the JWST were being paid for by sacrificing other programs within NASA's Science Mission Directorate (SMD). Initially estimated to cost $1 billion, JWST was now expected to cost $4.5 billion by the time it was launched in 2013. In addition, the Mars Science Laboratory (MSL) rover was $400 million over budget, and its launch would have to be delayed by 2 years. NASA's Edward Weiler, the Associate Administrator in charge of the SMD, simply did not have enough funds to keep everything going, and the President's 2010 budget request did not offer to help him out. With this preliminary budget request for FY 2010, it appeared that things were only going to get worse for the search for life beyond Earth.

But how much worse could it get? NASA did not yet have a new Administrator, who could argue with the Obama Administration about what NASA's budget and priorities should be. The First Dog, Bo Obama, a Portuguese water dog, was appointed prior to the naming of the next NASA Administrator, one measure of the likely importance of NASA in the new Administration. Former NASA Administrator Michael Griffin accepted a position as a professor at the University of Alabama, Huntsville, close to NASA's Marshall Space Flight Center. For Griffin, it was time to get out of Dodge.

Preparing for Battle: The main players in the U.S. exoplanet community gathered in Pasadena's Hilton Hotel on April 21–23, 2009, to plot a strategy for missions to be proposed to the Astro 2010 Decadal Survey. I chaired the opening session where three independent cost estimates for finishing SIM-Lite were presented: one JPL team estimated that another $1.15 billion would be needed, whereas a second JPL team came up with $1.41 billion. The estimate by the Aerospace Corporation was the highest of all: $1.65 billion for SIM-Lite and $1.4 billion for SIM-PH. Ouch. These estimates all placed SIM-Lite in the "flagship" mission category, concepts with total costs exceeding $1 billion. The competition was expected to be fierce for flagship missions. In spite of the high cost estimates, the consensus of the meeting was that SIM-Lite should be proposed and cast as the top priority of the exoplanet community, though it could hardly be said that SIM-Lite would be the only mission proposed to Astro 2010.

The 3-day-long meeting had presentations from dozens of enthusiasts with concepts ranging from as small as the Transiting Exoplanet Survey Satellite (TESS), with four 10-cm (4-inch)-diameter telescopes, to as large as a 16-m behemoth that might cost $7 billion or more. Several concepts involved flying a "star shade" along with a space telescope, where the star shade would serve as an external coronagraph, blocking the light from a host star so that the light from its planetary system could be seen just beyond the edge of the star shade. Although basically constructed of seemingly inexpensive black plastic, the estimates for a star shade of the proper diameter, about 50 m, still ran to $3.5 billion or more. TPF-C would likely be presented, in a 3.5-m by 8-m elliptical primary mirror format, along with a number of mini-TPF-C concepts with smaller-diameter mirrors. Astro 2010 would have the unenviable task of sorting through all of these competing exoplanet proposals and comparing them to equally compelling (and expensive) proposals for non-exoplanet astrophysics.

Overeating in Irvine: One of the primary dangers of serving on an NAS committee is the overabundance of food that is supplied during the committee deliberations, especially those held at the NAS Beckman Center in Irvine, California. Several excellent meals a day are served buffet style, along with morning and afternoon snacks. NAS committees are like armies; they travel on their stomachs. Many scientists react like the starving graduate students they once were and consume as much as they can. I fall in that unfortunate category and inevitably gain weight at such meetings.

The Astro 2010 Decadal Survey got underway with an all-hands "Jamboree" on May 11, 2009. As we sat in the auditorium that morning, filled with an ample breakfast but already anticipating the delights of the forthcoming luncheon, we watched live video of the HST servicing mission (SM4) lifting off from the Cape at 11:04 AM Pacific time, heading out to repair and replace several of Hubble's cameras and other critical parts. The NAS crowd watched the launch sequence in hushed silence, then burst into applause once the Space Shuttle *Atlantis* cleared the pad and soared into the sky.

Astronaut John Grunsfeld was now on his way to repairing the HST one more time. Grunsfeld was literally a "Hubble-hugger," full of enthusiasm for this pioneering space telescope. This would be his third trip to work on Hubble. HST's orbit is 300 miles above the Earth, 80 miles higher than the International Space Station, and has a different orbital inclination than the Station. These two factors meant that should *Atlantis* have any problems, the seven astronauts aboard would not be able to seek shelter in the Space Station. The backup plan for any emergency was the Shuttle *Endeavor*, which was sitting on a second launch pad at the Kennedy Space Center, ready to go if needed for a rescue. The loss of Shuttle *Columbia* upon return to Earth in 2003 was still fresh in everyone's mind, so NASA had to be prepared for anything, even a rescue in space.

HST had been in orbit for 19 years so far. With the latest repairs, involving five spacewalks, NASA planned to keep HST running for another 5 to 10 years. The successful launch of this final HST servicing mission was a relief to all in the Beckman Center audience: not only would the HST soon be capable of operating for years to come, but the enormous standing army responsible for planning and executing the Hubble servicing missions, and their sizeable annual budgets, could now be diverted to work on other high-priority NASA missions. SM4 alone had cost $1.1 billion, raising the total cost of the HST to date to about $10 billion. The purpose of the Decadal Survey was to decide just exactly what those future high-priority missions should be.

Survey Chair Roger Blandford started off the meeting by noting that the four Program Prioritization Panels would be independent National Research Council of the NAS (NRC) committees, charged with writing their own final reports, which need not agree with the overall summary report to be written by the Decadal Survey Committee itself. Yes, but would any report ever be read besides that of the Survey Committee's report? The Executive Summary of the Survey Committee's report would likely be the only product of Astro 2010 that would be studied in any detail.

We learned the names of the members of the various panels: of the 17 EOS panel members, I could count a total of only 4 people who could be counted as being interested in exoplanets, as evidenced by their past publication records. We 4 would need to stand united in order to have any hope of having an exoplanet mission come out on top, or at least close to the top. The NAS folks charged with determining the membership of the panels had done their job: their goal was to pick a group that had "all the skills and knowledge needed" and sufficient diversity "to ensure objectivity and impartiality." Once you choose the panel, you choose the panel's outcome, but what would that be?

We also learned that the NAS had received 108 responses to its "Request for Information" (RFI) call, soliciting relatively brief, 20-page white papers from the U.S. astronomy and astrophysics community for their ideas to be considered by Astro 2010. 58 of these RFI responses fell into the EOS panel's arena, over half of all of the relevant responses received. The 58 RFI responses had been divided up among the EOS panel members before the first meeting, and each of us had written a short summary of the ones we were assigned, noting whether any further information would be needed from their authors before we could finish discussing and ranking them all. Of the 58 RFI responses, 40 consisted of full mission proposals, with a total estimated cost of about $50 billion. Of the 40, 33 would eventually end up being seriously considered by the EOS panel, with 14 of those 33 proposals having something to do with exoplanets: 8 exoplanet proposals went for the brass ring and were designed to search for extrasolar Earths. We would have our work cut out for us—but first, it was time for lunch.

$750 Gets You the NASA Administrator's Job: The Associated Press announced on May 24, 2009, that Charlie Bolden had been offered and accepted Michael Griffin's former job as head of NASA. The Air Force General rumored to be considered for the job had turned it down on April 29, but the Marine Corps Major General had accepted. After all, how could Bolden, a former astronaut himself with four space flights to his credit, turn down the opportunity to lead NASA? Bolden had even been on the Shuttle flight that launched the HST into orbit back in 1990.

Oddly enough, the article mentioned that Bolden and his wife had "donated $750 to the Obama campaign in 2008." Was this considered full disclosure, or what? Michael Griffin may have been kicking himself for not having donated enough to *both* of the leading Presidential contenders in 2008 to ensure continued occupancy of the corner office on the 9th floor of NASA HQ. Would $800 have done the job, or $900 maybe?

$4 Billion Is All You Get: The Astro 2010 EOS panel met for a second time at the Pasadena Convention Center on June 8–11, 2009. The President's budget rollout for fiscal years 2010 through 2023 was presented to the EOS panel by Jon Morse, director of the Astrophysics Division at NASA HQ. Morse had some encouraging news for the EOS panel: the expectation was that the APD would have roughly $4 billion to spend on new space telescopes in the time interval of 2010–2020, assuming a flatline annual budget of $1.6 billion throughout the period. If Congress or the President insisted on spending even more on the APD than that, so much the better: but we could not count on more than $4 billion.

That $4 billion sounded like a lot, but the total cost of the HST to date had been over $12 billion, and JWST was now expected to cost at least $5 billion. In addition, Morse was fretting over a proposal, rumored to have been instigated by Maryland Senator Barbara Mikulski, a strong supporter of Hubble, to lay plans for yet another

HST servicing mission, to be called SM5, even though for historical reasons, SM5 would be the sixth HST servicing mission. The SM5 proposal had even been submitted to the EOS panel for our consideration. If the SM5 idea survived, the standing army of Hubble servicing mission engineers would remain on the APD payroll, at the expense of starting something new. Thankfully, the EOS panel would not have to deal with this particular hot potato, fully loaded with Maryland crab cake, as a few weeks later NASA would send out an email stating that the idea was "on hold" and that the associated future events were canceled, thereby saving Morse $20 million in the FY 2010 budget. SM5 had disappeared as quickly as it had appeared.

The $4 billion also paled in comparison to the optimistic wish list that had been submitted to our EOS panel, with over $40 billion of imaginative concepts for future missions still being considered. That meant the EOS panel would have to pick roughly the top 10% of the ideas submitted; 90% would fail to make the cut. The process was going to be brutal, but scientists are used to such low success rates in the annual competitions for federal research dollars. In addition, the other space-based Program Prioritization Panel would want a slice of the $4 billion pie, most likely to search for the long-hypothesized gravitational waves that Europe's Laser Interferometer Space Antenna mission might detect. LISA would involve multiple free-flying spacecraft connected by laser interferometers, and that mission would not be cheap to develop and fly either.

Still, $4 billion was a considerable sum, and the EOS panel members all began mentally deciding how they would spend the funds if given the chance. Given that the JPL's TPF-C concept would cost at least $4 billion, it was clear to me that TPF-C was out of the running. My personal mental ruminations thus focused on the pros and cons of how to rank SIM-Lite versus a so-called "probe-class" mission. Several concepts for such scaled-down versions of TPF-C had been submitted to EOS, with costs under $1 billion, cheaper than SIM-Lite's price tag of about $1.6 billion. These probe-class telescopes were not big enough to be able to image Earth-like planets, but they would take a giant step forward toward that ultimate goal. SIM-Lite could not image planets at all, but it would detect and determine the masses of the Earth-like planets orbiting the closest 100 or so stars, leading the way to a future TPF-C or TPF-I. But which step should we take next: SIM-Lite or a probe-class telescope? Astrometry versus direct imaging was shaping up as the main battle in deciding how I and the other exoplanet fanatics on the EOS panel would vote when the time came. As I sat at the familiar United gate 72 at LAX, waiting to board my flight home, I wrote out a detailed list of the pros and cons for SIM-Lite compared to a probe-class direct-imaging mission. There were good arguments to be made for both options, with no clear winner.

The flight back to Washington from Pasadena was delayed an hour and a half, so we arrived at Dulles at 1:40 AM. Dan Goldin, the former NASA Administrator who had championed the TPFs, was in Economy Plus on the same United flight, seated in the dreaded middle seat, dressed in a black polo shirt and blue jeans, but he had a limo driver waiting patiently for him in the Dulles baggage claim area. I thought about mentioning to Goldin that given the limited amount that could be spent on a flagship mission, TPF-C was likely out of the running for at least the next decade, but he was busy talking on his cell phone. Furthermore, the EOS panel deliberations were strictly confidential at this early phase of the NAS's secretive Survey process. I walked silently past Goldin and into the darkness of the night.

History Repeats Itself: TESS failed to make the cut for being considered for support by NASA's Explorer Program on June 19, 2009. TESS was the only exoplanet mission in the competition for NASA support as a Small Explorer (SMEX) mission. Even with a relatively low budget, compared to the figures being tossed around in the EOS panel's deliberations, TESS was not selected. The principal investigator, George Ricker of MIT, would have to continue to work on the concept, without guaranteed NASA support. Bill Borucki had had a similar difficult journey in getting the Kepler transit mission accepted for development, failing many times in successive competitions for final NASA approval. Evidently Ricker was facing the same uphill battle.

Kepler was taking data but had not yet proven the value of space-based transit photometry—perhaps NASA was not willing to start work on a second exoplanet transit detection mission until the first one had proved its mettle? Ricker had attended the open sessions at the Pasadena EOS panel meeting earlier in the month and had told me over breakfast one morning some good news about Kepler. Contrary to claims that had been made by the CoRoT team, Kepler was finding that most solar-type, G dwarf stars are as quiet as the Sun, making Kepler's job of differentiating between stellar noise and exoplanet transit events that much easier than it might be otherwise. Ricker's TESS proposal could only be strengthened by the success of the forerunner Kepler Mission.

Ninth Floor of 300 E Street SW Gets a New Occupant: Incoming NASA Administrator Charlie Bolden faced his Senate confirmation hearing on July 8, 2009, answering questions in a packed room on Capitol Hill. Bolden's answers and comments made it clear that planetary science and astrophysics would be taking the back seat at NASA compared to increased funding for Earth science during the Obama Administration. Bolden stated that "we have to look at Earth, our planet, and NASA has to lead." No mention was made by Bolden of NASA's exploration of the Solar System or the universe beyond. Bolden was confirmed on July 15. Evidently climate change was in, and exploration was out.

Showtime in the Webb Auditorium at NASA HQ—I: The same day that Bolden was facing the Senate, the Kepler Program Scientist at NASA HQ inquired by email if I would be willing to participate in the first major press conference about Kepler's results. The role would be a familiar one for me at press events at NASA HQ, to provide the "Big Picture" context of Kepler's exoplanet discoveries. Political pundits have talked about how former President Bill Clinton loved to discuss his

Administration's policies and initiatives. Clinton was well versed in the fine details that are usually only known and understood by staffers. President and former actor Ronald Reagan, on the other hand, who often spoke from handheld note cards, was charitably known as a "Big Picture" guy. Hence I was being offered the chance to play the role of Ronald Reagan at the first Kepler press event. I immediately accepted the invitation, noting that by chance I had given the weekly DTM seminar that morning about exoplanet discoveries, a talk that extolled the virtues of the Kepler Mission. I was primed and ready to go.

The heavily scripted press event, with rehearsals and considerable preparatory discussions, went so far as to include a request from Ed Weiler, SMD's chief, that we do not wear coats or ties, a request that was easy enough to honor for a press conference that was held on the afternoon of August 6, 2009, in steamy Washington, DC. Jon Morse led off in the Webb Auditorium, literally setting the stage, followed by Bill Borucki, who showed Kepler's first light image and then the new science result, about to be published in *Science.* Sara Seager and I followed with our own Big Picture perspectives. My final point was that Bill and the Kepler team were about to determine just how crowded the universe was with potentially habitable worlds, a heady vision indeed.

The press event went well, and although Kepler was not quite ready to announce the detection of any new exoplanets this early in the game, Bill Borucki did prove the astonishingly high photometric precision of Kepler's giant camera: the light curve for a previously known, transiting exoplanet in the Kepler target field of Cygnus. This hot Jupiter, called HAT-P-7b (see Figure 3.1), the seventh exoplanet found by the ground-based Hungarian Automated Telescope (HAT) transit search, was used by Kepler as a test case for how much better a space-based telescope could follow a transit, compared to a ground-based telescope, buried within the flickering, dense atmosphere of Earth. Kepler's HAT-P-7b data proved not only that Kepler had the photometric precision and stability to be able to detect Earth-size planets, but also showed the slight dip in total light when the hot Jupiter passed behind the host star, as well as the brightening in reflected light as the fully illuminated side of the gas giant began to be pointed toward Kepler, just prior to its complete disappearance behind the star. While CoRoT had been able to demonstrate these same stunning physical effects for its first two hot Jupiters, CoRoT-1b and CoRoT-2b, the fact that Kepler was now also taking superb data meant that CoRoT had serious competition.

Showtime in the Webb Auditorium at NASA HQ—II: This time the big press conference was about the complete success of the SM4 mission to refurbish Hubble. The astronauts who did the work joined the press conference on September 9, 2009, where a gorgeous image of the planetary nebula NGC 6302, the Butterfly Nebula, was featured, thanks to the newly installed Wide Field Camera 3. Everything was working again now on Hubble, and Ed Weiler predicted that the HST was good for "at least five years, perhaps as much as ten, with a big emphasis on perhaps." SM5 was ruled out though—there would be no more repair missions for Hubble. That

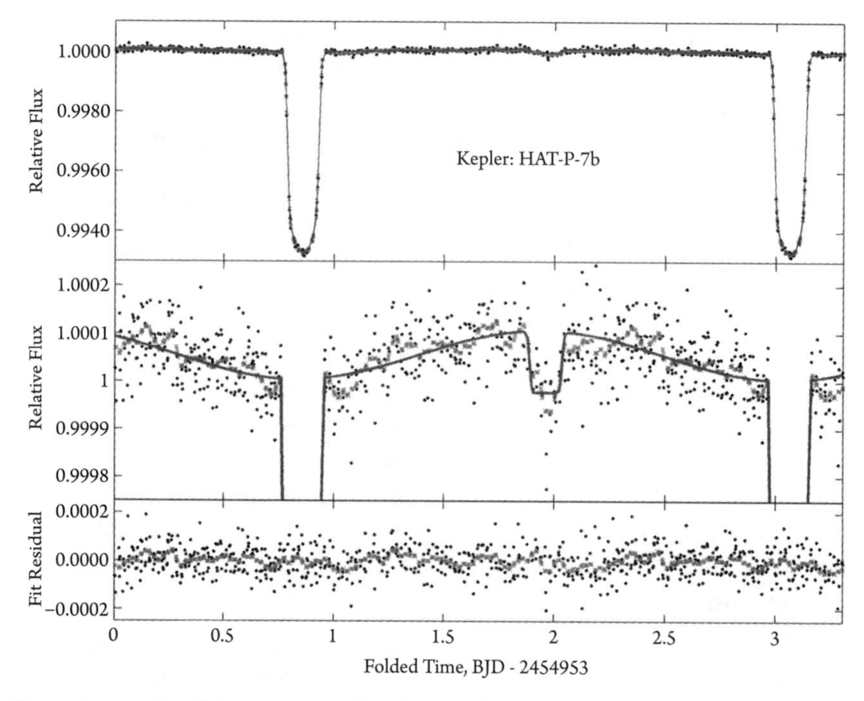

Figure 3.1 Kepler's light curve of the HAT-P-7b hot Jupiter, proving the required photometric precision and showing the brightening due to the light from the star-facing side of the exoplanet just before and after the secondary eclipse (occultation).

meant that the EOS panel could continue to plan on spending $4 billion in 2010–2020, give or take a billion.

Maryland Senator Barbara Mikulski had the honor of presenting the new Hubble images in her role as the "Godmother of Hubble." The STScI, which operates Hubble, and the Goddard Space Flight Center (GSFC), which built Hubble, are both located in Maryland; and Senator Mikulski had given strong support to Hubble and its repair missions through her role as the chairwoman of the Senate Appropriations Subcommittee that funds NASA. Mikulski referred to Hubble as "the people's telescope," and Hubble was now as good new, actually better than new, considering the new technology incorporated in Hubble's new and refurbished cameras.

4

Barcelona Is in Catalonia, Not in Spain

"If we knew what it was we were doing, it would not be called research, would it?"

—Albert Einstein (1879–1955)

Much of the world's exoplanet community gathered in Barcelona to assess the status of our efforts and to plan the way forward. The meeting was held on September 14–18, 2009, with the seductive title "Pathways Towards Habitable Planets." I chaired the session on the morning of September 16 where a major milestone was announced during a talk by Daniel Rouan of the Paris Observatory.

CoRoT-7b was declared by Rouan to be the *first detection of a rocky planet,* based on a refined estimate of its mass, 4.8 times that of the Earth, obtained by the unprecedented investment of 70 hours of Doppler monitoring by the Swiss High Accuracy Radial velocity Planet Searcher (HARPS) spectrometer, located deep in a basement below the 3.6-m telescope at ESO's La Silla Observatory in Chile. Doppler spectroscopy had led to the discovery of the first confirmable planet around a sun-like star, 51 Peg b, in 1995 by Michel Mayor and Didier Queloz of the Swiss Geneva Observatory; and HARPS had been developed and built by Mayor and his team. Queloz led the team that used HARPS to measure the mass of CoRoT-7b, whose radius of 1.7 times that of the Earth had been measured during 153 transits observed by the CoRoT space telescope. That size and mass meant that the mean density of CoRoT-7b was 5.6 g per cubic centimeter, essentially the same as that of the Earth, 5.5 g per cubic centimeter. CoRoT-7b thus appeared to be made of silicate rock and iron metal, just like the Earth.

But that was not all Rouan had to announce that morning. The HARPS Doppler data showed that CoRoT-7 was orbited by a second planet as well, now dubbed CoRoT-7c, with a mass 8.4 times that of Earth. CoRoT-7c did not show up in the transit data, but the painstaking Doppler observations had found it. The orbital periods of these two exoplanets were 0.85 days and 3.7 days for b and c, respectively, meaning that they orbited so close to their sun-like star that they were both hot super-Earths, with surfaces more similar to a volcanic lava lake than a hospitable ocean planet. CoRoT had found two hot super-Earths for the price of one.

The auditorium in the CosmoCaixa science museum in Barcelona responded to Rouan's talk with enthusiastic applause for both the CoRoT and HARPS team. The announcement attracted worldwide attention, and the story was emailed to others on Yahoo more often that day than the second most popular story, which involved "candid thoughts" by President Obama regarding the popular celebrity Kanye West. Even the politically oriented Huffington Post put the story on their home web page. Exoplanets were hot news that day, even uninhabitable hot super-Earths.

European scientists reported on the status of their future space planning efforts, termed Cosmic Vision, the European equivalent of the Decadal Survey underway in the United States. Sadly, the European version of TPF-I, the Darwin mission, had not been chosen for further study, as it was judged to be insufficiently mature in terms of the new technology it would require. Only one exoplanet mission remained in the running, a medium-class mission called PLAnetary Transits and Oscillations of stars, or PLATO. PLATO would also be a transit mission: did the world really need another transit mission? The TESS proposal was being patched up and would be reproposed to NASA at the next opportunity: given that TESS would find transit exoplanets for JWST follow-up, it seemed that TESS would eventually be accepted.

Plans were underway for the E-ELT to achieve "first light" in 2018, with a spectrometer capable of detecting the tiny Doppler wobble (9 cm per second) caused by Earth-like exoplanets, and with enough resolving power to image Earth-like planets around the closest stars. ESO's E-ELT was billed as being the world's strongest exoplanet program. Provided that Kepler showed that Earth-like planets were commonplace and hence could be expected to be found around the Sun's closest neighbors, E-ELT would take it from there.

Another speaker stressed the fact that stellar noise would complicate the detection of exoEarths by Doppler spectroscopy. Geneva's Francesco Pepe made the point that having a spectrograph marginally capable of measuring a 9-cm-per-second wobble would not be good enough, especially if more than one exoplanet orbited in the system, as was to be expected. The highest-precision Doppler spectroscopy needs quiet stars, but most of the 100 stars closest to Earth are low-mass, red dwarf stars, which are notoriously noisy due to star spots and other phenomena. Hence even the E-ELT's best spectrometer would not be able to determine the masses of the Earth-like planets around the 100 closest stars: their host stars would be too noisy to permit this intrusion into their privacy. For sun-like stars, the stellar noise was about 10 times worse for Doppler detections than for astrometric detections. That meant that we needed SIM, or SIM-Lite, in order to determine their masses, the basic property that cannot be determined definitively by direct-imaging detections.

The CoRoT-7b announcement had made the case for needing to know the mass of an exoplanet before deciding if it was a rocky world or not. The ESA planned to launch in 2012 the Gaia space astrometric telescope, which would be able to determine the masses of thousands of nearby Jupiter-mass exoplanets, but only SIM could do Earths.

We heard that the SIM project at JPL was on hold, as NASA HQ was waiting to hear from Astro 2010 before committing any more funds to SIM. A satellite meeting on Doppler versus astrometric planet detection reached a group decision that astrometry was better at finding the nearby Earths than Doppler spectroscopy, though the latter should still be pursued. In the final session, the conference summary speaker, Pierre Lena of the Paris Observatory, said that SIM-Lite should be flown before moving on to direct-imaging missions, even those of probe class. SIM-Lite would find the Earth-mass targets for the direct-imaging missions to follow. Lena concluded with the idea that the ESA should consider partnering with the United States on the SIM-Lite concept and then continue this partnership with a joint Darwin/TPF-I mission. Sharing costs is an excellent way to make an expensive space telescope seem more affordable; this arrangement had worked in the past with Hubble and was also underway for JWST.

Lena called for ESA and ESO to unite in making plans for Europe. Evidently the ground-based astronomers in Europe were planning on having the E-ELT make space-based exoplanet telescopes unnecessary, consistent with the failure of ESA's Cosmic Visions to include anything beyond another transit mission. The ESO mantra appeared to be *if it can be done from the ground, it will be done from the ground, first.* It was hard to argue with this logic, given the outcome of Cosmic Vision, what was happening to the NASA Astrophysics budget, and to the EOS panel's upcoming painful version of Sophie's Choice.

Third Time's the Charm: The third and final EOS panel meeting was held at the NAS Keck Center in Washington, DC, on September 23–25, 2009. Jon Morse led off by telling us that the "blue lake" of funding available for future NASA APD missions was beginning to drain away: his estimate of $4 billion at the June EOS meeting might have to be lowered to about $2.3 billion, just 3 months later. If this was correct, at this rate, the blue lake would disappear before the end of 2009. The main leak was projected to be cost overruns for JWST. We also learned that the particles and gravitational waves Program Prioritization Panel had settled on LISA as their top mission, at a total cost of $2.3 billion. Even with European partnership on LISA, that did not leave much for the EOS panel, should LISA be ranked highly by the Survey Committee, which oversaw both of our efforts.

This meeting represented the end game for the EOS panel. While a dozen concepts had survived scrutiny to this stage regarding their science merits, likely costs, technical readiness, and risks, there were four main contenders for the top ranking. One was the International X-ray Observatory (IXO), with a total cost of perhaps $5 billion, to be shared with ESA. The IXO proponents considered IXO to be next in line for a top ranking, given the outcome of past Surveys and the need for program balance, but it looked like IXO could bust our now smaller budget. A second concept was basically a replacement for Hubble, another 2.5-m optical and ultraviolet space telescope that had the virtue of being proven technology with a relatively modest cost estimate of $1.1 billion, though the Aerospace Corporation team charged with assessing costs and risks had not yet studied this concept. A third was a concept born from the merger of two EOS proposals, one searching for dark energy, and one looking for exoplanets by searching for microlensing events (see Figure 4.1). Although the science was very different, both proposals called for a similar, modest-sized (1.1 m or 1.5 m) wide-field, infrared telescope. The EOS merged concept was called the Wide-Field InfraRed Survey Telescope (WFIRST), a clever name that played on the fact that it was intended to beat a competing ESA mission, Euclid, to the goal of determining the "w" parameter that is used to characterize the amount of dark energy in the universe. WFIRST was intended to be the first to measure "w." Measuring "w" was also the top priority of WFIRST, meaning that if push came to shove during the development of WFIRST, which always seemed to happen with state-of-the-art space telescopes, and something had to be dropped in order to cut costs, the microlensing planet hunt was likely to suffer. WFIRST's total cost estimates came in at about $1.6 billion.

The fourth concept was a generic true exoplanet mission, with both SIM-Lite and a probe-class imager being the leading contenders, though a 4-m space telescope coupled with a star shade was also still in the running, in spite of a cost likely to be over $5 billion. SIM-Lite looked downright affordable in comparison: with $600 million already invested, SIM-Lite needed just over $1.4 billion to complete, though the Aerospace estimate came in at about $1.8 billion. The probe-class concept was less than $1 billion, by definition.

We exoplanet enthusiasts presented the cases for the exoplanet missions still in the running. It was clear that such a mission would meet many of the science goals that had been outlined for Astro 2010 by the separate science panels charged with assessing the state of astronomy and astrophysics in 2009. I presented the long list of pros and cons for SIM-Lite and the probe-class imager that I had started while sitting in LAX after the Pasadena EOS panel meeting and then summarized the

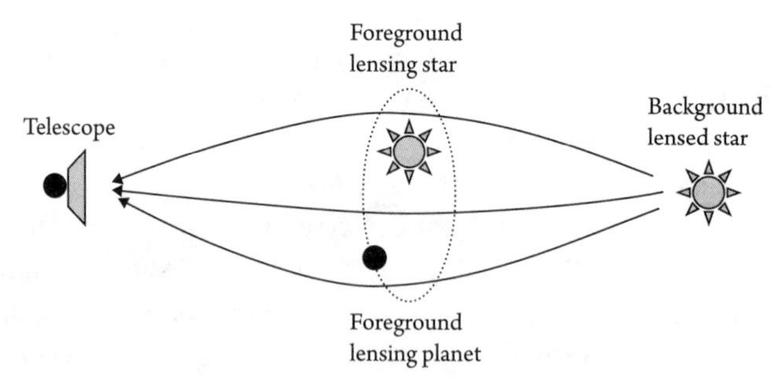

Figure 4.1 During a gravitational microlensing event, light from a background star is bent toward the Earth by the gravitational pull from an unseen foreground star and planet, resulting in an apparent brightening of the background star for a period as long as a few months.

sense of the Barcelona meeting that SIM-Lite should be flown prior to a probe-class direct imager.

Then it was time for a vote, disarmingly termed a "straw vote." The first vote came out with WFIRST on top, followed by SIM-Lite, IXO, a probe-class exoplanet imager, and the Hubble clone. A second vote was limited to just these top five, and resulted in IXO jumping ahead of SIM-Lite, and the Hubble clone ahead of the exoplanet imager. Considering that the scores on that second vote were so close for IXO (7.87) and SIM-Lite (7.83), a runoff, third vote was held for IXO versus SIM-Lite, which IXO won by a vote of 13.5 to 10.5. The fractional tallies resulted from the straw vote procedure of each EOS panel member giving 1 point to their first-place choice, 0.5 points for second place, and so on. These three straw votes quickly settled into a final vote that was as hard as concrete: the EOS priority ranking was WFIRST, IXO, and SIM-Lite, respectively, and that was that.

The total cost assumptions behind this voting were that WFIRST would run $1.5 billion, IXO would cost the United States $2 billion, and SIM-Lite needed about $2 billion. That ran up the total tab for APD to about $5.5 billion, without counting a possible LISA mission. At the time, I considered it a victory that SIM-Lite came out as well as it did during the straw votes, given that only 4 EOS panel members out of the 17 voting could be considered exoplanet nuts, and we 4 were split between SIM-Lite and the direct imager as our first choice. APD would need a bigger blue lake than Morse was projecting if SIM-Lite was to survive.

As chair of the EOS panel, my colleague Alan Dressler of the Carnegie Observatories gave the summary of our deliberations to the Decadal Survey Committee a week later, on October 3, 2009. Dressler noted that if Morse's worst-case estimate was only $2.3 billion to spend, then only WFIRST could fly in the 2010s, with IXO getting a start later in the decade. If the full $4 billion was available, then an exoplanet mission could be started late in the decade as well. The EOS preference was for that exoplanet mission to be SIM-Lite, but Dressler recommended that the choice between SIM-Lite and a direct imager be made by an expert panel at some later date, when funds became available. It was now up to the Survey Committee to consider the recommendations from all four Program Prioritization Panels and decide the final rankings for Astro 2010.

Given that roughly three-quarters of the EOS panel work in the areas of cosmology and extragalactic astronomy, where dark energy rules, those few of us who cared about exoplanets would have to be satisfied with a bit of microlensing exoplanet hunting carried piggyback on WFIRST, as well as a third-place finish for the generic exoplanet mission. *Once you choose the committee members, you choose the committee's outcome.* This adage was holding true for the EOS panel. But as long as NASA Astrophysics had $4 billion to spend in the next decade, I hoped that the exoplanet community would get something significant out of the Astro 2010 rankings other than a microlensing survey.

Special CoRoT Issue of *Astronomy & Astrophysics* (A&A): The new results on CoRoT-7b were published in the leading European astrophysical journal in a special issue released on October 25, 2009, covering the early results from CoRoT. However, the published papers told a somewhat different story from what we had applauded in Barcelona, where the claim was made that the first rocky planet had been found. Once the A&A referees had finished with their reviews, one of the two published papers was not quite so strident about their claims. Queloz's paper stood by the mass estimate for CoRoT-7b of 4.8 Earth masses, noting that this meant it might be "made of rocks like the Earth, or a mix of water ice and rocks." The other paper simply stated that the mass of CoRoT-7b was less than 21 times that of the Earth and that as a result, the planet's composition "remains loosely constrained without a precise mass." So what was CoRoT-7b? A rocky super-Earth, or a melted ice giant, a water world? No definitive answer could be given.

A few days later, we learned that the French space agency, along with its partners, had agreed to extend the operations of CoRoT for several more years, until April 2013, a half year after Kepler was slated to finish its prime mission in October 2012. CoRoT was not about to give up the game earlier than Kepler.

Bill Borucki sent out an email to the Kepler science team on October 31, 2009, complaining that a recent article in *Nature* had made it sound like Kepler would have a hard time finding Earth-sized planets because of a known problem with the electronic hardware on board the spacecraft. Borucki pointed out that a mitigation strategy for this problem was already underway and that he expected Kepler to be able to fully achieve the desired photometric accuracy: Kepler's results on HAT-P-7b had already proven this to be true. The *Nature* article was entitled "Planet hunt delayed," which was not true and cast Kepler as being in a "fierce race" with CoRoT and with ground-based exoplanet searches. That much *was* true.

Loose Lips Sink Flagships: Someone appeared to leak highly confidential news about what was being considered by the EOS panel. A November 19, 2009, story on SPACE.com stated that ESA's Euclid Mission, one of the contenders for a slot as a large-class mission in Europe's Cosmic Vision competition, was considering adding in a microlensing, planet-hunting element along with its prime goal of being a "dark energy" mission, a search for one of the apparently missing major components of the universe. The news story noted that a similar merging of exoplanet and dark energy, arguably the two hottest topics in astrophysics, was being considered in the United States. What in the United States could that be, besides WFIRST? Yes, indeed, such a merger had been considered by the EOS panel and presented to the Survey Committee, but that was not supposed to be revealed. I sent an email with a link to the news story to Dressler, the chair of the EOS panel, and he sent back a reply wondering if Astro 2010 had a leak, or was this just a coincidence? The idea of combining these two hot topics had a natural appeal, as both required a similar space telescope, a relatively modest-sized telescope operating at optical and near-infrared wavelengths, capable of wide-field, high-resolution imaging. It made perfect sense to combine both science goals on the same mission, and an earlier version of Euclid had made that same case, so maybe it was just a coincidence. Or maybe it was not.

Andrew Carnegie Meets Johannes Kepler: Bill Borucki's invited plenary talk to the American Astronomical Society (AAS) on January 4, 2010, was given to a ballroom packed with several thousand astronomers, where Bill presented the first five exoplanet discoveries made by Kepler (see Figure 4.2), using just the first 43 days of observations. Four were hot Jupiters, but one was a hot super-Earth, dubbed Kepler-4b. At my invitation, the Kepler Mission held a science team meeting after the AAS meeting at the Carnegie Institution's administrative headquarters in Washington, DC, at 16th and P Streets, within sight of the White House. The Carnegie Institution Trustees had overridden the advice of the generous benefactor we sometimes call "Uncle Andrew" who had declared that once an institution begins to build fine buildings for itself, dry rot would begin to set in. The Trustees constructed an outstanding example of what can be done with marble and stone

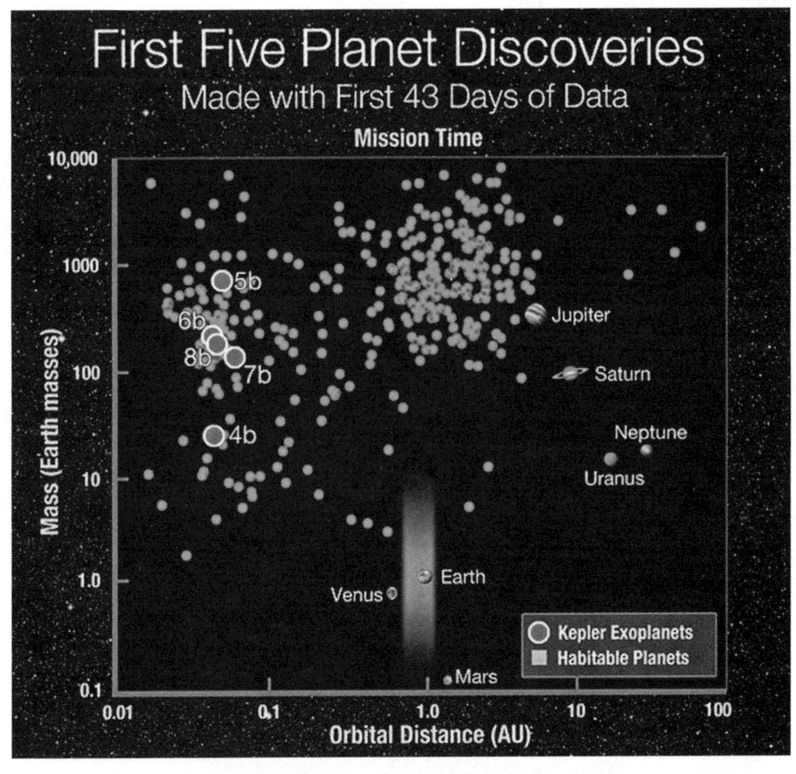

Figure 4.2 The first five exoplanets discovered by the Kepler Mission and confirmed by Doppler spectroscopy, showing their masses and orbital distances from their stars, compared to the Solar System's planets and exoplanets discovered by ground-based transit surveys (Courtesy of NASA).

when one wants to make an impression in a city with no shortage of impressive buildings and monuments.

The P Street location was eminently suitable for a meeting of the equally eminent Kepler science team. We learned that Kepler had uncovered over 200 planet transit candidates in the first 4 months of data collection, all with orbital periods of less than 30 days. Eight stars appeared to have at least two transiting planets, and five showed evidence for at least three transiting planets. That was truly amazing: Kepler was finding compact planetary systems, where multiple planets were jammed together on orbits closer to their stars than that of Mercury, the innermost planet in our Solar System, all confined to orbits in a plane thin enough that when viewed edge-on by Kepler, one planet after another would proceed to block out some of the host star's light, predictably, over and over again. There were even indications of 53 hot super-Earth candidates in the data taken so far.

The Kepler harvest was well underway, and the team turned to plans for the harvest festivals, namely, making sure that the Kepler results were featured at every upcoming scientific conference we could manage, from the AAS meetings, to the more general meetings of the American Association for the Advancement of Science (AAAS), where thousands of scientists of all flavors would meet to sample the harvest fruits.

A few days later, an article appeared in the *Astrophysical Journal* (ApJ) by Steven Vogt of the Lick Observatory (see Figure 4.3), my DTM colleague Paul Butler, and their Doppler planet search team announcing the discovery of the first super-Earth orbiting a sun-like, G dwarf star. In fact, their 16 years of observations implied that the star 61 Virginis hosted at least three exoplanets, with minimum masses of 5.1,

Figure 4.3 Steven S. Vogt, the designer of the HIRES spectroscope on the Keck I telescope (Courtesy of DTM, Carnegie Institution for Science).

18, and 24 times that of the Earth. The paper wisely concluded that this discovery "portends that . . . Kepler and CoRoT will find many multi-transiting systems." David Charbonneau of the Harvard-Smithsonian Center for Astrophysics (CfA), who found the first transiting exoplanet, HD 209458 b, in 1999 (see TCU), and his colleagues had published in late 2009 the ground-based detection of a transiting super-Earth, 6.6 times as massive as Earth, orbiting the red dwarf star GJ1241. The implications of both of these discoveries could not be better for the Kepler Mission and the search for life beyond Earth.

We wanted to trumpet the incredible discoveries that Kepler was beginning to make, not only to the American public, whose taxpayers had invested over $600 million in making the Kepler Mission a reality, but also to our fellow astronomers and astrophysicists who we knew would be gathering in early 2012 to consider whether the Mission should be given an extension beyond the prime mission life of 3.5 years. The 2012 Senior Review would make the decision about whether Kepler would have to close up shop or continue finding more planets, planets with the longer and longer orbital periods that could only be discovered with an extended mission phase. A meeting in DC in early 2010 was not too early to begin plotting our strategy for that critical Senior Review. I took on the task of getting Bill Borucki invited to give a plenary talk at the 2011 AAAS annual meeting, along with a special session on Kepler results, a task that I was well equipped to perform as the incoming chair of the Astronomy Section of the AAAS. It helps to be an insider when you want to get things done.

President Obama's first full budget request for NASA came out on February 1, 2010, with the expected focus on climate change research, and not a word about any telescopes beyond mentioning the successor to Hubble. JWST, the NASA Astrophysics Division's pride and joy, did not appear by name in the funding highlights of the budget plan for fiscal year 2011 and beyond.

Keeping the Faith: The SIM Science Team, a competitively selected group who had spent several years working on and planning the full-scale "SIM-classic" mission design, held a telephone conference call to update the team about recent events at NASA HQ and in Europe on February 11, 2010. The SIM project leader at JPL optimistically told the group that if SIM was awarded a top ranking in the Astro 2010 Decadal Survey, then SIM would be able to proceed with a launch during the 2010 decade: SIM's technology was sufficiently developed to permit such a relatively early launch date. Another JPL scientist reported that Europeans were considering collaborating with NASA on SIM.

Because the EOS report was still under embargo, I had to keep quiet about the true status of SIM's chances in the Decadal Survey, which were slim to none, to even come in with a third-place ranking. First place was out of the question. If there were any leaks about the EOS panel rankings, the leaks had evidently not reached JPL. Worse yet, the EOS panel had recently learned that the $4 billion we were planning on spending for NASA had dwindled to about $2 billion. The blue lake of future mission funds shown on NASA's projected budget charts was beginning to drain away, and the $2 billion figure was expected to drop further. At $2 billion or less, even the $700 million that EOS was bookkeeping for a generic exoplanet mission would completely disappear. In fact, that figure meant that there might not be any new flagship mission in the 2010 decade.

I had been participating in planning for the SIM mission since 1996, and SIM was dead, though the SIM team did not yet know it. By the NAS rules, I could not tell them, nor did I wish to speak the awful truth.

The likely fate of SIM was soon revealed by *Nature* in the February 18, 2010, issue. *Nature* published the results of the rankings of a group of seven leading U.S. astronomers who were not involved in the Decadal Survey process and hence were free to speak their minds. SIM did not make their list of the top seven projects for the ground and space. The top-ranked space telescope was a generic TPF, oddly enough, though the group agreed that the total price tag for any TPF would have to decrease significantly below that of TPF-C. Odder still, the Joint Dark Energy Mission (JDEM) came in ranked as the *fourth priority* for a space telescope. JDEM had been the primary driver for the creation of the WFIRST mission concept by the EOS panel, coupled with the Microlensing Planet Finder (MPF) mission. MPF did not figure at all into the rankings of the *Nature* group. Clearly this group of seven accomplished astronomers and astrophysicists had arrived at a ranking for future NASA space telescopes that was diametrically opposed to what the EOS panel had decided on, a ranking that was likely to be echoed in the final Decadal Survey report. *Once you choose the committee members, you choose the committee's outcome.* If you choose another group of committee members, and rerun the test, you may very well get a different rank-ordered list. *Nature* had inadvertently proven the hypothesis—Q.E.D.: *quod erat demonstrandum*, the mathematicians' equivalent of calling out "checkmate" once a theorem is proven.

First Imaging of Exoplanets Takes the Cake: At the annual AAAS meeting at the waterfront convention center in San Diego on February 20, 2010, two teams were jointly awarded the society's Newcomb Cleveland Prize for the best paper published in *Science* magazine in 2009. Christian Marois and his co-discoverers of the HR 8799 system of visible exoplanets and Paul Kalas and his Fomalhaut detection team won the coveted award. The award of this AAAS prize implicitly recognized the importance of what a TPF could accomplish for imaging exoEarths, even if the Decadal Survey was unlikely to do so.

While sitting in the San Diego airport, waiting for my flight back to Dulles, I opened an email from Jon Morse, head of NASA's Astrophysics Division. Jon wanted to speak to me as soon as possible, so I called him on my cell phone and learned that he wanted to know right then and there if I was willing to take over the chair of the NASA Advisory Council's (NAC) Astrophysics Subcommittee (APS). The NAC is the one group formally chartered by Congress to give "advice" to NASA. I had been on many NASA committees before but never before one that was allowed to "advise" NASA, much less chair the committee that held sway over all of NASA's existing space telescopes, from Hubble to Kepler, as well as telescopes in development, such as JWST, and the future telescopes being considered by the ongoing Decadal Survey. I could not resist accepting Morse's request on the spot. Morse was pleased, and immediately hung up, without providing any further details. I did not know what I would be in for, never having even attended a meeting of the APS, but if NASA wanted advice from me, they were going to get it.

Naming Rights Are Not Cheap: The Kavli Institute for Theoretical Physics (KITP) at UC Santa Barbara (UCSB) held a conference from March 29 to April 2, 2010, under the rubric "Exoplanets Rising." While the ITP part of KITP had been founded in 1979 by the NSF at UCSB and supported ever since, Norwegian-born businessman Fred Kavli had purchased the naming right for the ITP for a mere $7.5 million, about the same amount as is necessary to get a professional sports stadium named after you. Kavli went on to support numerous other scientific institutes and to establish in 2005 a biannual $1 million Kavli Prize in three areas overlooked by Sweden's Alfred Nobel—one being astrophysics. Sweden's Holger Crafoord had established the $0.7 million Crafoord Prize in 1980, also in areas not covered by the Nobel Prize such as astronomy. Evidently Scandinavian rivalry is intense to this day, to the lasting benefit of astronomy and astrophysics.

Drake Deming (then at the NASA GSFC) spoke at the KITP conference about just what JWST would do for the field of exoplanet science. Assuming that the TESS mission proposal was accepted and that TESS was able to fly before JWST, TESS would deliver a target list of nearby, bright, red dwarf stars with transiting exoplanets, perfect for further study by JWST. JWST would not be able to image these worlds, as its coronagraph would perform far worse than those being built for use on the ground, but it could perform the same transit spectroscopy that had been perfected by Spitzer and Hubble, and could do it much better as well. Assuming that each low-mass, M dwarf star has one super-Earth, Deming estimated that the combination of TESS and JWST would be able to perform a detailed study of between one and four habitable super-Earths. JWST would be able to determine the temperatures of their atmospheres, and look for water and carbon dioxide. Between one and four? If Kepler found a lower frequency of habitable super-Earths than unity, the tally might drop to zero.

April Showers: While visiting Cape Canaveral on April 15, 2010, President Obama announced his new vision for the nation's space program. NASA would first send astronauts to a nearby asteroid and then on to the planet Mars. Mars made sense as a suitably ambitious, long-term goal for human space flight, and in fact had been the ultimate goal advanced for NASA in President George W. Bush's 2004 VSE plan. But a trip to an asteroid was a new idea; the VSE had instead directed NASA to stop by the Moon again on the way to Mars. Although a clear vision statement for NASA's future was a welcome move forward, the Obama vision did not make any mention of space telescopes capable of studying Earth-like planets, unlike Bush's VSE. I had learned at one of the EOS panel meetings that the words read off the teleprompter by Bush during the 2004 VSE announcement calling for such a space telescope had been written by the NAS staffer running our panel. Evidently the Obama Administration was content to restrict the search for life beyond Earth to Mars.

My first meeting as chair of the NAC Astrophysics Subcommittee occurred on April 12, 2010, primarily to consider the data policy for the Kepler Mission. NASA HQ, and many astronomers not on the Kepler team, wished for all the data from Kepler to be released to the public once the data pipeline had finished removing artifacts and systematic errors; but the Kepler team made the argument that until the team could consider all 3.5 years of data, there was a good chance that over-eager astronomers sifting through only a portion of the data would start making false claims about what had been found. In order to preserve the integrity of the Mission, Bill Borucki asked that the team be able to keep private the data of the 500 best exoplanet candidates for the full 3.5 years of the prime mission. As a Kepler Science Team member, I recused myself from the discussion. The APS settled on a compromise close to Bill's request: 400 targets for about 1.5 years. It would be up to Jon Morse to decide the issue.

The backup charts in Morse's presentation to the APS included the traditional budget overview, showing that APD had been cut by about $200 million between FY 2009 and FY 2010. Worse yet, the final chart made it clear that APD had no more than about $2.3 billion to spend on Astro 2010's recommendations, as we on the EOS had feared.

A few weeks later, I attended my first NAC Science Committee meeting at GSFC on April 21, 2010. This was the week of the 20th anniversary of the launch of HST, and SMD head Ed Weiler crowed that after the successful SM4 repair mission, Hubble was in good enough shape to last for another 20 years. HST now had new gyroscopes, new batteries, and new solar panels, all for the bargain basement total price to date of $18 billion. This total included a prorated share of the cost of the six Space Shuttle flights that were used to launch and service Hubble. While Hubble could thus look forward to several more decades of spectacular astronomical imaging and spectra, Weiler wondered how much longer the astronomical community would be willing to pay the tab for keeping a basically 1980s-vintage, though recently refurbished, space telescope operational.

Even without any more servicing missions in the pipeline, the Hubble support bill came to about $100 million a year, enough to start serious work on a new, 2010-vintage space telescope. Hubble would have to be safely de-orbited once it outlived its usefulness, as its massive primary mirror would create quite a crater if it hit the land, or a harmless splash if it hit the ocean. NASA was planning a specialized mission to de-orbit HST after 2020. The last servicing mission, SM4, had installed a docking ring that could be used for that final HST mission, which might cost $500 million and change.

At the GSFC meeting, I asked Ed what he thought about the lack of any mention of TPF-like telescopes in the new Obama vision for NASA, and he was similarly chagrined. Ed said he thought looking for habitable planets was the most important goal of all for NASA and that he had gone into astronomy in order to look for life beyond Earth. Ed pointed out that it was up to the Decadal Survey to make the case for imaging habitable worlds. I kept my lips tight, but it made no difference: the TPF Flagship, along with SIM-Lite, had already been torpedoed and sunk by a combination of NASA's dwindling Astrophysics Division budget and the dark energy priorities of the EOS panel.

Later that same week, ESO announced that the E-ELT would be built in Chile's Atacama Desert, a high, dry, cloudless region perfect for telescopes. The press release stated that the E-ELT would be able to image exoplanets after it began operating in 2018 on Cerro Armazones in northern Chile. The cost was estimated to be over $1 billion. The GMT was also planned for northern Chile, while the TMT would go on Mauna Kea, Hawaii, in part to please private donors who insisted on a clear U.S. title to the land on which TMT would stand.

Something Smells Fishy: A JWST review board raised serious issues with the project, prompting the project manager to announce on May 14, 2010, that the launch would be delayed to October 2014 in order to perform additional testing. Even with 40% of the APD budget, JWST needed more funds, as well as additional time, or else more schedule slippage would occur, further driving up the cost beyond the estimate of $5 billion.

A billion here, a billion there, and pretty soon you are talking real money, as Senator Everett Dirksen was reputed to have remarked about the federal budget.

Double or Nothing: On June 15, 2010, the Kepler Mission released to the public the data from its first 4 months of observations, with the data on 400 stars being held back for further confirmation. With various refinements in the data pipeline, the claim made at the Carnegie Institution science team meeting in January for having found over 200 planet transit candidates in this same data set had now grown to 750 planet candidates. This was more exoplanets than had been found in the previous 15 years of ground- and space-based searching. Kepler had more than doubled the total number of candidate exoplanets with just a few months of observations, and still had at least 3 more years of data to process. Kepler was in the process of revolutionizing the field of extrasolar planets, rewriting the textbooks and further galvanizing the search for exoEarths.

Meanwhile, the Astro 2010 process was winding down, and the results were to be revealed to NASA HQ in August 2012. Jon Morse had tasked a group of exoplanet enthusiasts to put together an action plan in case Astro 2010 should come out strongly in favor of exoplanets. Did even Morse not know what was coming down the pike?

The JPL folks continued to work on SIM technology development, and the SIM science team continued to hold its breath. JPL planned a live broadcast of the Astro 2010 unveiling in the von Karman auditorium on the JPL campus. It had been announced that the Astro 2010 Decadal Survey was entitled "New Worlds, New Horizons in Astronomy and Astrophysics" (NWNH), with the first phrase giving hope that exoplanets had come out first in the Survey rankings. Perhaps SIM-Lite was on top? Evidently no one at JPL knew what was really coming their way in the form of the Astro 2010 rankings. There appeared to be no serious leaks about the EOS panel rankings, much less those of the Survey Committee. The live broadcast would be a shock to many at JPL and around the world.

Who Did This?: While I was observing at Carnegie's Las Campanas Observatory (LCO) in Chile on June 28, 2010, President Obama's National Space Policy was released. Amazingly, buried deep on page 12 was a direct order to the NASA Administrator to "search for planetary bodies and Earth-like planets in orbit around other stars." I could not have said it better myself. This was very much like the directive given by George W. Bush in the 2004 VSE. While not mentioned during the Cape Canaveral speech in April, this wording was clearly all the exoplanet community could hope for in terms of Presidential support.

Who had managed to get this wording inserted in the new Space Policy this time? Charlie Bolden? Ed Weiler? Jon Morse? Somebody at the NAS? We could only hope that this time the Presidential mandate would come along with the funds necessary to perform the mandate, as the APD was fresh out. The 2004 VSE call had turned out to be largely an unfunded mandate. In fact, after an initial sharp rise, the annual funding for SIM and the TPFs had dropped precipitously. There was a higher-priority, competing item in the APD at that time: JWST. But JWST was now expected to be launched in 2014, and thereafter some of the JWST cash flow could be diverted to an exoplanet flagship, with the President's blessing, right?

Wrong Again: The NAC Science Committee held a meeting at NASA HQ on July 13–14, 2010. Before the meeting began, Ed Weiler handed me a copy of *Space News* with a nice color photo of Jon Morse above the fold on the front page, but with a not-so-nice headline reading "Webb Telescope Cost Growth Prompts Mikulski Demand for Outside Review." Maryland's Senator Barbara Mikulski, champion of HST, GSFC, the STScI, and all others things Maryland, was worried about JWST's continual cost and schedule overruns. JWST was now consuming at least $444 million each year and asking for more, with its launch delayed to mid-2014. Ed Weiler was tired of being pestered for more funds for JWST. Mikulski called for NASA to establish an independent review team to figure out what the problem was with JWST. Her June 29, 2010, letter to Charlie Bolden gave him 30 days to appoint the review team, which was to report directly to the Administrator.

Morse was quoted in the *Space News* article as saying that APD was already in the process of forming a separate JWST review team, the Test Assessment Team (TAT), to be led by JPL's John Casani. As it turned out, Casani would also lead the review team that would play the role demanded by Senator Mikulski. As the chair of the Senate Appropriations Subcommittee with jurisdiction over NASA's budget, when Senator Mikulski asked for something from NASA, she was used to getting it. Building JWST was going to cost more than the $5 billion or so planned so far, but how much more than that could it cost? And when would it launch?

5

Witness Protection Program

First get your facts; and then you can distort them at your leisure.
—Mark Twain, 1835–1910

The Astro 2010 Decadal Survey was released on Friday the 13th, August 2010, an entirely appropriate choice for the release date. I decided to flee DC and spend some time on an island on Florida's Gulf Coast while the Decadal Survey hit the fan. I took a long bicycle ride at the same time that the Survey was being revealed back in DC and thought about what was happening around the country, as the few winners, and many more losers, were tallying the outcomes of Astro 2010. I did not want to even think about what the mood would be like in JPL's von Karman auditorium.

Astro 2010 made it likely that it would be at least another decade before NASA could launch a true exoplanet mission, like Kepler, or a SIM-Lite, or a TPF-Lite. Kepler would be the only game in town for quite a while: TPF-T was it for the foreseeable future. In fact, Ed Weiler had asked the Kepler team to see if it could announce a habitable zone exoEarth by the end of 2011: Ed wanted to make sure that Kepler found the first habitable world, perhaps around a red dwarf. That would be good enough to start Kepler's march to glory.

The SIM science team held a post-mortem telecon a few days later, on August 19, 2010. JPL Director Charles Elachi pointed out that while SIM's low ranking was disappointing, the SIM team should still support the Astro 2010 report, as it did call for significant spending ($100 million to $200 million) on exoplanet technology development efforts for the next decade. On a more practical note, a SIM program manager said that although there was as yet no direct word from NASA HQ about the SIM project's fate, it would be a good idea to send in to JPL any bills for work performed to date. The unopened boxes of glossy booklets extolling how SIM would discover and weigh the closest Earth-like planets would be unceremoniously tossed into the JPL recycling bins. The SIM project was shutting down after investing $600 million in developing microarcsecond-level astrometric technology that might never be used. The JPL exoplanet crowd would have to wait yet another decade and see what Astro 2020 might recommend.

In comparison to the gloom at JPL, the dark energy folks at NASA HQ were ecstatic and immediately began making plans for WFIRST.

Win a Few, Lose a Few: A flurry of exoplanet discoveries and retractions emerged in the shadow of the Astro 2010 release. Reanalysis of the CoRoT-7b Doppler data by several teams led to mass estimates ranging from about 3 to 9 Earth masses, and cast doubt on the very existence of CoRoT-7c, which soon thereafter disappeared from consideration. However, Christophe Lovis of the Geneva Observatory presented Doppler evidence for the existence of an astounding 7 exoplanets all in orbit around the same star, HD 10180, based on six years of HARPS Doppler monitoring. 5 of the 7 appeared to be fairly massive super-Earths, with the other 2 not having well-defined masses, though one might have as little as 1.4 times the mass of the Earth.

The same day, August 25, 2010, Borucki and Matt Holman of the Harvard-Smithsonian Center for Astrophysics held a press conference to announce the detection of at least two, and possibly three, hot exoplanets orbiting the star Kepler-9. The Kepler transit data showed the existence of two Saturn-mass exoplanets, with their masses derived by the extent to which their mutual gravitational interactions affected their individual orbits, leading to small changes in the times when they transited Kepler-9. Besides the two Saturns, there was transit evidence for a super-Earth with a mass in the range of one to seven times that of Earth, but the evidence was so sketchy that this third body was not yet considered worthy of the appellation Kepler-9d. If real, it would be the smallest exoplanet found to date, with a radius about 1.5 times that of Earth. The next day, the Kepler team was alerted that a vote would be taken soon on whether to designate this "object of interest" as Kepler-9d or not.

The Kepler team evidently wished to avoid the fate of CoRoT-7c, as well as the lingering uncertainty about the mass of CoRoT-7b. A democratic vote would decide the issue, at least for the moment. Borucki sent an email to the team on August 30 confirming that the team had voted unanimously in favor of Kepler-9d. The regiment voted.

Tar and Feathers?: The first Town Hall community meeting to explain the rationale behind the Astro 2010 rankings was held at DTM in northwest Washington, DC, on September 2, 2010. I had volunteered the location and to serve as the local host, and the NAS folks had gladly accepted the offer. Scott Tremaine, a Decadal Survey member, had taken the train down from Princeton that morning to serve as the guinea pig for gauging the astronomical community's reaction to Astro 2010 at a personal level. Scott gave a detailed overview of the entire Decadal Survey process, and of the rankings for ground and space, and then fielded questions for another hour. The questioners were civil, and Scott handled their barbs and queries equally civilly. One attendee pointed out that Astro 2010 was really effectively Astro 2020, given that JWST was spending all the funds available in the 2010s, and asked if the Survey folks knew this. The response was that the Survey was asked not to worry about JWST. Another questioned why NASA Astrophysics was being asked to support the dark energy search of WFIRST: wasn't that the purview of physicists and the Department of Energy, which had pushed for JDEM? In fact, a *Science* article in early 2011 would list WFIRST as being in the domain of U.S. particle physics, based on its focus on dark energy.

Apparently the nearly month-long time period between the release of Astro 2010 and the first Town Hall meeting was sufficient to let any overheated tempers cool down. Scott was pleased that the presentation went so well, in spite of giving it at DTM, which he considered to be an exoplanet hotbed. Clearly, the Decadal Survey folks were worried that the microlensing planet hunt portion of the top-ranked WFIRST space mission would not be sufficient to gain support from the exoplanet community, who had asked for, and expected, much more. The Survey claim was that dark energy and microlensing planet searching were co-equal science goals for WFIRST; but given that few exoplanet devotees worked on microlensing, a feeling of dissatisfaction with Astro 2010 remained.

But Green Bank Has No Cell Phone Service: While attending the 50th year anniversary celebration of Frank Drake's Project Ozma search for radio signals from extraterrestrial civilizations using one of the Green Bank radio telescopes, I had a confidential telephone call with Jon Morse on September 12, 2010. A landline had to be used, as Green Bank, West Virginia, is a radio-quiet zone—even microwave ovens are prohibited—in order to prevent stray electromagnetic emissions from interfering with the work of the Green Bank Observatory's sensitive antennas. Oddly enough, the 26-m radio telescope Drake had used for Project Ozma was named the Tatel Telescope, after the DTM physicist, Howard E. Tatel, who had designed it in the 1950s.

Morse was concerned about the Astro 2010 top recommendation of WFIRST. The Europeans were considering the Euclid mission as their next medium-class space telescope, along with the planetary transit mission PLATO. Euclid was now back to being a pure dark energy mission and would be in direct competition with WFIRST. Euclid was planned for launch in 2018. Given the situation, ESA wanted to know if NASA would like to collaborate with ESA on Euclid, as they had done with HST and were doing with JWST. Morse estimated that WFIRST could not launch until 2022, giving Euclid 4 or more years to skim the cream off of the dark energy milk pail, leaving only some skim milk for WFIRST. A similar threat had been posed by CoRoT's launch over 2 years before Kepler, although in that case it was clear that it was Kepler that was skimming the cream, not CoRoT. Morse was willing to see if NASA would buy into Euclid at a level large enough to ensure that some of the U.S. astronomers who were clamoring for dark energy results could be part of the Euclid science team.

Morse was also concerned about how to handle JWST's continual demands for more funds: should he give the project an extra $100 million next year or not?

As the new chair of the APS, Morse's advisory committee, Jon wanted to let me know about what was going to happen at the APS meeting scheduled for a few days later, September 16–17, 2010. In a private meeting with Jon and Ed Weiler just before that APS meeting began, the problem with Euclid participation became clear: some U.S. astronomers wanted to keep NASA completely out of the Euclid mission. Euclid needed U.S. detector technology, and without a NASA partnership, that sensitive technology might not be granted to our European allies. These astronomers wanted to ensure that WFIRST would do a much better job at measuring "w" than Euclid would. Euclid would be merely a pathfinder for WFIRST. Those who intended to be the leaders of WFIRST apparently were hoping that a successful WFIRST mission would give them a chance to shake the hand of the King of Sweden, right after picking up the Nobel Prizes that were sure to follow.

Roger Blandford presented the Astro 2010 results to the APS meeting, noting that WFIRST was expected to cost $1.6 billion, and that if the United States should agree to collaborate with ESA on a joint dark energy mission, Astro 2010 insisted that the United States should play the leading role.

As expected, the question of whether NASA should partner with ESA on Euclid became the most contentious issue during the APS meeting and forced me to have the APS take a formal vote about how to proceed, rather than reach a consensus that we could all agree on. The vote was 10 to 2 in favor of keeping open the option of a partnership with ESA on Euclid. A second vote was held to decide the level of NASA participation, with one abstention: a 20% share in Euclid won with 7 votes, while the 33% share offered by ESA and suggested by Morse received 4 votes: 20% meant no change in the presumed NASA share in Euclid.

We seemed to have solved the Euclid problem, at least for the moment, but I had stunned the crowded room into silence earlier in the meeting when JWST was being discussed. My question to all in the room of how much more than $5 billion JWST would cost was met with absolute silence. How could we give advice to NASA if we had no good idea of what the APD would have left in the checking account after JWST?

There was no easy answer. I adjourned the standing-room-only meeting on time, left NASA HQ, and headed back up Rock Creek Park from the Mall to my lair at DTM in northwest DC, wondering how much the final bill for JWST would be. The Casani report, the work of the Independent Comprehensive Review Panel (ICRP), about the JWST situation was scheduled to be delivered to Senator Mikulski on October 1, and that report would give us all an answer, though it may not be one we wanted to hear.

Please Turn Down Your Stereo: The Kepler Science Team met at NASA Ames on September 20–21, 2010, where Ames Center Director Pete Worden declared that Kepler was Ames's "flagship mission." The spacecraft was functioning well and was expected to last another 10 years, provided that NASA supported an extended mission beyond the initial 3.5 years. Natalie Batalha (see Figure 5.1) presented a top-secret result: Kepler had found its first rocky planet, Kepler-10b, a hot super-Earth 37% larger than Earth, with a mass almost 5 times as large. The press release about

Figure 5.1 Natalie Batalha, one of the science team leaders of NASA's Kepler Mission (Courtesy of NASA).

Kepler-10b would appear in a month or two, and we had to keep quiet until then. When the paper appeared in early November, the radius would be refined to be 42% larger than that of Earth, making it the smallest transiting exoplanet found to date. With a mean density of 8.8 g per cubic centimeter, 60% higher than that of Earth, it appeared that Kepler-10b was a rocky world, perhaps with a bit more iron in its core than Earth. The orbital period was only 20 hours, though, so this was a really hot super-Earth, orbiting at a distance barely above the surface of its star. Kepler-10b would be a molten lava world, inhospitable to life.

So far, so good. However, we also learned some disquieting news. The processed Kepler data was not meeting the maximum noise level specified in the Mission's requirements document. In fact, the noise was about 50% worse than expected, and that meant that Kepler might not be able to detect Earth-like planets: Earth-size objects with 1-year orbital periods around sun-like stars. The Mission had a tiger team working on this critical problem, but time was running out before some data had to be released. Whereas the source of this unexpected noise was unclear, there was a hint that the problem was not with Kepler's CCD detectors and electronics but with the target stars. When the data was processed for the best subset of stars, the noise was at the expected level, based on what we knew about how noisy our Sun is—implying that the true problem was that most stars are noisier than the Sun, something that only Kepler could have determined. Although that meant that Nature had thrown us a curve ball, it was expected that NASA could not blame the Kepler Mission for not thinking that the Sun was an exceptionally quiet neighbor.

My Planet Is More Habitable Than Your Planet: A major brouhaha broke out over the bragging rights for discovering the first habitable planet on September 29,

2010. A team of Swiss and European scientists had previously used ground-based Doppler spectroscopy to discover evidence for at least four planets in orbit around the red dwarf star Gliese 581. Two of them, Gliese 581c and Gliese 581d, were super-Earths that orbited right about at the inner and outer edges of the star's habitable zone (HZ), respectively, where liquid water might be stable at the planet's surface. Although close, it was not quite clear if either of these two worlds were truly in the HZ, as the definition of the HZ depends on uncertain assumptions, such as the atmospheric conditions on the exoplanet.

My DTM colleague Paul Butler (see Figure 5.2), and Lick Observatory astronomer Steven Vogt, were also following the bumps and grinds of Gliese 581 with Vogt's High Resolution Echelle Spectrometer (HIRES) on one of the Keck telescopes in Hawaii. Vogt and Butler held an NSF press conference to announce the discovery of two more super-Earths in the same system, Gliese 581f and Gliese 581g. Whereas the former exoplanet orbited too far out to be habitable, the latter planet orbited right smack dab in the middle of the HZ. Gliese 581g appeared to be the best candidate of all for the first HZ super-Earth, with a mass between 3 and 4 times that of the Earth, orbiting its red dwarf star every 37 days.

Finding HZ, Earth-like planets was supposed to be Kepler's job; but ground-based telescopes appeared to have won this race, even if they had cheated by looking around low-mass, red dwarf stars, where HZ planets are much easier to detect, compared to more massive stars like the Sun. *If it can be done from the ground, it will be done from the ground, first.*

The news of the first habitable super-Earth swept around the Internet, and the emails starting coming in. One hyperbolic Internet story quoted an astronomer

Figure 5.2 R. Paul Butler, a pioneer of Doppler spectroscopic exoplanet detection (Courtesy of DTM, Carnegie Institution for Science).

who said "odds of life on newfound Earth-size planet '100 percent.'" Fox News insisted on having me down for a live interview the next day, September 30, 2010, and sent out a limousine to whisk me downtown to their studios, across from Union Station and the Capitol Building. The interview would be for the Shepard Smith show on Fox News National, one of the more balanced news shows on a network not necessarily known for balanced coverage. I had tried to get Fox News to have Paul or Steve do the live interview instead, but the Fox News scheduler had been told that both of them were in transit elsewhere and unavailable. Steve was indeed on an airplane back to California following the NSF press conference in northern Virginia the day before, but Paul was sitting in his house in DC, not that far from the Fox News studios, waiting for a plumber to arrive and install a new hot water heater. Habitable planets are important, but being able to take a hot shower whenever you want to take one is important too.

Where Did They Go?: A few days later, I was expecting Jean Schneider to send out his usual email to the exoplanet world about the discovery of the two new Gliese 581 planets, but no such email arrived. Could the big news somehow not have made it across the pond to Europe? I pinged Butler and Vogt about this, and Steve guessed that perhaps Schneider wanted to give the Swiss-European team who found the first four Gliese 581 exoplanets a chance to see if they believed the claims for two more planets in the same system. Fitting Doppler data for multiple planets is as much an art as it is a science, and fitting the Gliese 581 data for six exoplanets was certainly pushing both the state of the art and the state of the science. The two new exoplanets did not even appear in the "Latest news" section of Schneider's *Extrasolar Planets Encyclopedia* web site.

What was going on? I soon found out at an exoplanet meeting held in Torino, Italy, shortly after the NSF press conference. Francesco Pepe, a member of the Swiss-European team that had found the first four planets for Gliese 581, gave a talk on October 11, 2010, where he presented their team's reanalysis of their Gliese 581 Doppler data and found no evidence for Butler and Vogt's new HZ planet Gliese 581g. Pepe's team was still able to fit their data with just four planets. Did this explain the absence of Gliese 581f and 581g from the *Encyclopedia*? Schneider eventually did list these two planets on his influential web site, but only in a special category, reserved for exoplanets that are "unconfirmed, controversial, or retracted." It was hard to say if this was better than not being listed at all in the *Encyclopedia*. By October 12, even that listing disappeared from the web pages, with no explanation given. Pepe's talk in Torino seemed to have convinced Schneider that the claims for Gliese 581f and 581g were spurious, but it was clear that this argument had not been settled by Pepe's talk. It appeared that the main discrepancy was that when two very different approaches were used to analyze the same complicated data set, two very different results could be obtained. But which one was correct? "We need more data" is the refrain always heard from astronomers in similar situations.

Steve and Paul's ApJ article noted that if their Doppler planet search effort had been able to detect a HZ Earth-mass exoplanet orbiting such a nearby star—Gliese 581 is a little less than 20 light-years (about 6 parsecs) away—that implied that HZ exoEarths must be commonplace. Given the small number of nearby stars that had been scrutinized, they estimated that Eta_Earth, the fraction of stars with potentially habitable Earth-size planets, might be as high as "a few tens of percents." Determining the value of Eta_Earth was the main scientific goal of the Kepler Mission; and while the Gliese 581 detection could not yield the statistically sound determination that Kepler would provide, this Doppler discovery buoyed the hope that Kepler's Eta_Earth would indeed be close to unity.

Another Kind of Desert: Theoretical models of the formation of planets by the core accretion mechanism used by several groups predicted that there should be a "planet desert" for exoplanets with masses in the range of a few to several tens of Earth masses, that is, the super-Earths, with short orbital periods, less than about 50 days. The main reasons for this theoretical prediction were twofold. First, once exoplanets grew to super-Earth masses, the models predicted that they could pull in hydrogen and helium gas from the surrounding protoplanetary disk and then continue to suck in enough disk gas to turn into gas giant planets with masses of several hundred Earth masses or more. Second, the models assumed that the rest of the super-Earths would migrate inward through interactions with the disk gas and end up stranded on orbits very close to their stars, turning them all into hot super-Earths. The net results of these two effects would be a paucity of warm and cool super-Earths. If true, this would mean that Kepler would find this exoplanet desert early in the mission, as the short-period transiting planets are the first ones to transit repeatedly, leading to a firm claim of detection. However, Kepler would only find the planet desert by not finding any short-period super-Earths other than hot super-Earths, whereas the goal was to find warm and cool super-Earths, and HZ Earths to boot. Confirming this predicted exoplanet desert would be a downer for the Kepler Mission: confirmation meant *not* finding exoplanets. This was not good.

One of the first hints that this theoretical prediction might be erroneous emerged from a talk by Geoff Marcy of UC Berkeley (UCB) at the Torino meeting on October 11, 2010. Marcy presented the results of a Doppler survey of 166 sun-like stars; and after making corrections for missing some detections, Marcy and his UCB colleague Andrew Howard had found evidence that smaller mass planets were more frequent that larger mass planets, for orbital periods of 50 days or less. In fact, they were strikingly more frequent: there appeared to be about 10 times as many super-Earths as more massive planets in their sample. Given this result, Marcy and Howard went on to extrapolate to longer-period orbits and claimed that their data implied that 23% of sun-like stars could harbor a short-period world with a mass between half and twice the mass of Earth. While not quite an estimate of Eta_Earth, the implications were positive for what Kepler would find: the Doppler data implied that Kepler would find that the predicted planetary desert would in fact be a planetary oasis.

Hurricane Season Arrives Late in mid-November: The long-awaited report requested by Senator Mikulski concerning the problems with JWST was released, at 5 PM on the Wednesday evening before the November 11, 2010—Veterans Day holiday, the perfect time to release a disturbing assessment. The usual assumption about such a release was that the bad news would be forgotten by the time folks went back to work the following Monday morning.

The independent review board, chaired by JPL's John Casani, had done a thorough job of studying what had happened to date with JWST's design, development, and building process and had a good idea of what else needed to be done to finish the job. The Casani report now estimated that it would cost at least $6.5 billion to build JWST and that the launch date would have to slip more than another year to the fall of 2015. To do that, JWST would need an extra $250 million in the current fiscal year (FY 2011) and in the year after. The report further suggested a shake-up of the project's management. NASA agreed and started by moving the project office from the GSFC to NASA HQ, where a closer watch could be maintained on the costs. The first task of the new JWST management team would be to develop a bottom-up budget estimate for finishing the telescope.

Under the usual NASA spending rules, firewalls were maintained between the Divisions in the Science Mission Directorate so that a cost overrun in the Planetary Sciences Division, for example, could not be paid for by removing funds from the APD. Assuming that the firewalls would remain in place, I did a quick mental calculation. Paying for the JWST cost overruns from the APD budget pie would mean that JWST would be eating 70% of the entire pie, leaving only about half as much for the rest of the APD: critical assets such as Hubble, Kepler, and a host of other space telescopes already in operation or development. A 50% cut to these existing assets was unthinkable, but that seemed to be what would be required to keep JWST going, if the Casani estimates were correct.

I followed the Casani report release from home, as I was headed to Dulles that evening for another observing run in Chile; but before I left, I sent a quick email reply to a reporter from the *New York Times* who had asked for my reaction to the Casani report. I emailed him my honest assessment: that if the firewalls held, and if the President and Congress were unwilling to provide new funds to support JWST, then JWST would leave nothing but devastation in the APD. Given the severity of the federal budget crisis, then in the throes of what was being called the *Great Recession*, and calls that day by conservative Congressmen for further federal budget cuts, this seemed to be the logical outcome of saving JWST.

Being used to talking to reporters, who appreciate short sound bites, I led off my email response with "This is NASA's Hurricane Katrina," referring to the disaster that had struck New Orleans in 2005 and the ineffective response of the Federal Emergency Management Agency (FEMA) in the hurricane's aftermath. I concluded by saying that JWST's cost overruns meant that there would be no funds available to start any new large missions in the current decade of the 2010s, and hence there was no need for NASA to have participated in the just-completed Astro 2010 Decadal Survey. WFIRST could not fly until about 7 years after JWST launched, and a 2015 launch for JWST meant at best a 2022 launch for WFIRST. There would be nothing new to launch in the decade of 2010.

As the Chair of the APS, I had effectively just given NASA some "advice," though it was not advice that they would want to hear. I grabbed my bags and headed out to Dulles for the long trip to the Las Campanas Observatory. It would be good to spend a week working in isolation on a Chilean mountaintop, looking for exoplanets, with a telescope and an instrument that did not exist only in PowerPoint presentations.

The Popular Press Decides the Issue: The December 5, 2010, issue of *Parade* magazine, an insert in many Sunday newspapers, ranked Vogt and Butler's announcement of the first habitable rocky world, Gliese 581g, as number 6 in its list of the most amazing discoveries of 2010. Gliese 581g beat out number 7, which was the discovery by archaeologists of the oldest leather shoe ever found, complete with shoelaces. The 5,500-year-old shoe was discovered in a cool, dry Armenian cave, a perfect environment for preservation. There was no response from the Europeans to the ranking for Gliese 581g—*Parade* magazine is not circulated in Europe. Let's just keep this to ourselves, okay?

NASA HQ agreed with *Parade* magazine. When Charlie Bolden released his list of NASA's major achievements for 2010, one of them was the announcement of Gliese 581g. While NSF had held the press conference for Gliese 581g, NASA claimed a share as well, as Vogt and Butler had used the Keck HIRES spectrometer built by Vogt to perform the Doppler measurements, and NASA owned one-sixth of the Keck Observatory. Paul wryly pointed out to me that he and Steve had been unable to get any of NASA's Keck time, which was annoying when NASA decided to take credit for their Gliese 581 work. NASA's HIRES Keck time was being devoted mostly to following up Kepler's transiting exoplanets, trying to get their masses, so Paul and Steve were out of luck.

Don't Shoot the Messenger—Part I: In the face of flat federal budgets, on December 6, 2010, President Obama requested a major boost in funding for an Earth-observing satellite being built by NASA on behalf of the NOAA, the National Oceanic and Atmospheric Administration. There was no mention, however, in the Administration's budgetary "anomaly request" of the "JWST anomaly" that had just been revealed by the Casani report.

A week later, an astronomer affiliated with JWST told me of a meeting she had had where an Office of Management and Budget (OMB) official had brought along a copy of the *New York Times* article with my Hurricane Katrina quote. Evidently the White House subscribed to the *Times*, even though it is published in the distant northeastern hinterland of New York, far outside the Washington Beltway, as the quote had caught the attention of President Obama. Considering the unfortunately premature congratulations by President George W. Bush of his FEMA Director for having done a "good job" in New Orleans, a job that was in reality almost as much of a disaster for New Orleans as Katrina was, and a job that led to the FEMA Director's resignation, President Obama was undoubtedly sensitive to a reference to Katrina in the context of any federal agency.

The JWST astronomer implied that my quote might lead to the diversion of NASA funds away from space science and into Earth science, and she asked that the APS keep its eye on the ball and prevent any such dire budgetary action. In particular, she requested that the APS focus on the science drivers for JWST and accept the cost consequences, with the eventual reward being not only the discoveries that JWST would make but the funds that would flow to the JWST user community for supporting their research costs, the same business model that helped to make Hubble such a success. The STScI had pioneered one-stop shopping: HST telescope time was awarded by the orbit, along with most of the funds need to perform the research. One wondered at times if some astronomers wanted the research funds more than the HST orbits.

In the wake of the Casani report, which had been undertaken in a bit of a senatorially induced rush, the newly reorganized JWST project office at NASA HQ had begun a serious, prolonged evaluation of just what JWST needed to be completed, starting with a blank budgetary spreadsheet. The idea was to make sure that there would be no more surprises like the Casani report. Senator Mikulski and her staff would be watching.

Any Questions?: The newly revamped JWST project office held a Town Hall meeting at the Seattle AAS meeting on the evening of January 10, 2011. The session featured free food, thanks to the corporate support of Northrop Grumman Space Technology (NGST), the prime contractor for JWST, which was itself originally abbreviated as NGST: the Next Generation Space Telescope. Evidently the space industry was running short on acronyms and would have to repurpose a few. NGST later renamed itself to Northrop Grumman Aerospace Systems (NGAS), a dubious improvement for an acronym implying intestinal distress.

JWST might be costing billions, but at least we picked up some chicken teriyaki sticks at the Town Hall, though it was questionable if they could be referred to as "free food." NASA's Eric Smith, the deputy lead, had the dubious honor of describing the JWST replan. Eric outlined the plan to start with a blank spreadsheet and figure out exactly what was remaining to be accomplished in building and testing JWST. Once the replan was accomplished, the results would be presented to Charlie Bolden, then to the White House Office of Science and Technology Policy (OSTP), OMB, and the Congress, and then to NASA's advisory bodies, such as the APS. Finally, the replan would be made public. Until then, mum's the word.

Mum's the word characterized the standing-room-only audience's reaction to the presentation. When Eric finished, and there was a call for questions from the audience, not a single question was raised. The eerie silence was deafening—no one had the temerity to question what would happen to future missions like WFIRST, much less what had already happened to the TPFs and SIM-Lite. JWST was the third rail of the APD—touch it only at your peril.

At an Exoplanet Exploration Program Analysis Group (ExoPAG) meeting in Seattle a few days before, I had asked an old friend, a European astrophysicist, what had been the reaction in European circles to the dismaying news about JWST's cost overruns; and he chuckled that the European response was "snickers," and he was not referring to candy bars. The Europeans had their own problems with funding flagship-class missions: the L-class mission queue was stalled, and only M-class missions like PLATO and Euclid still had a chance.

The brightest spot of the AAS meeting was the announcement by Natalie Batalha of the discovery of Kepler-10b. A few weeks later, NASA Ames's Jack Lissauer had a paper in *Nature* describing Kepler's discovery of a truly astounding planetary system, Kepler-11, with six exoplanets orbiting in a plane thin enough that all six produced transits when viewed from Kepler. Kepler was still NASA's primary exoplanet asset, and given the JWST situation, it was looking like that might remain to be the case for at least a decade to come.

Flagships for the Star Fleet: The President's budget plan for FY 2012 was rolled out on schedule on Valentine's Day, February 14, 2011. The first bullet in the NASA "Funding Highlights" summary stated that the agency's budget request was the same as for FY 2010. Given the overall federal budget situation, with an annual budget deficit of $1.5 trillion, NASA was lucky to not be cut, as was being proposed for most other federal agencies. In the words of OMB, *"flat" was the new "up"*—a flat budget was the best that an agency could hope for. However, compared to the current year's budget, FY 2011, NASA would be cut by 6.6% in FY 2012, if the President's budget request was accepted by Congress. The budget table evidently told a different story from that presented in the "Funding Highlights" summary.

SMD did considerably better than NASA overall: the SMD budget proposal for FY 2012 really was flat compared to FY 2011. Funds for JWST had now been broken out into a separate line item in the SMD budget, the better to allow close scrutiny of the project's costs. Still, the FY 2012 request for JWST was $354 million, compared to $439 million in FY 2010. It appeared that there would be no help in FY 2012 from the Administration in stopping the rising waters in the JWST project, though JWST was at least now separated from the APD. Or was it?

We learned more about NASA's plans at a meeting of the NAC Science Committee at NASA HQ on March 4, 2011. NASA officials said that they would have to drop the plans to develop the WFIRST flagship mission, the top space priority of Astro 2010. "Flagship" had become the new "f-word" at NASA. Instead, NASA hoped to collaborate with ESA on the Euclid dark energy mission. ESA had not yet made the final decision between flying Euclid or PLATO as its next medium-class mission, but NASA made it clear it was not interested in partnering on PLATO, only on Euclid. NASA was effectively putting its thumb on the scales in favor of Euclid, while the Europeans tried to weigh the advantages and disadvantages of these two space missions, which were effectively competing head to head against each other.

ESA responded that it would not consider collaborating with NASA on Euclid until after October 2011, when it would announce the winners of the next two M-class mission slots. If Euclid happened to be one of the two chosen, that would be the time to talk to NASA. ESA was trying to preserve the sanctity of its mission selection process and to keep NASA's fingers off the scales. In fact, ESA was leery of any further joint missions with NASA, given the budget problems at NASA, period. ESA directed the proponents of the three teams working on proposals for a large (L-class) mission to remove any NASA participation from their proposals. ESA planned to launch the L1 mission by 2020, and they decided that they could not count on NASA being able to participate on that time schedule. The ESA LISA and IXO study teams would have to go it alone; NASA's support for the U.S. LISA and IXO teams was canceled on April 5, 2011.

Charlie Bolden made an unscheduled visit to the March 4, 2011, NAC meeting. Normally the agenda is slavishly followed during meetings of the NAC, its committees, and subcommittees, as these public meetings are advertised in the Federal Record at least 30 days ahead of time, and people want to listen in or attend only when the items of interest to them are to be discussed. But when the Administrator decides to drop by for a talk, you toss the agenda aside and see how long you can keep him in the room. Charlie asked all the Science Committee members to say a few words about themselves; and when my turn came, I pointed out that I was working on looking for new planets beyond the Solar System. Charlie's eyes lit up, and he stopped the round from continuing by asking me what was new in the search for other worlds. I replied "Kepler," and then I described some of the hot new results that were being readied for publication. I could not resist throwing in another sound bite, saying that Kepler was delivering *flagship-quality science for the price of a Discovery-class mission*. That was a true statement: Kepler had cost about $600 million, but it was delivering results worthy of a multibillion-dollar TPF.

Charlie Decides the Issue: During testimony on April 11, 2011, before the Senate Appropriations Commerce, Justice, Science (CJS) Subcommittee, Charlie Bolden stated that JWST would likely not launch until 2018 rather than the planned 2014 date. The FY 2012 budget contained less than $400 million for JWST, and Bolden intended to continue spending on JWST at that rate until launch rather than try to find an extra $500 million a year for FY 2011 and FY 2012 in order to shoot for the 2015 launch date suggested by the Casani report. Charlie evidently decided to keep the firewalls in place and keep the JWST conflagration from spreading: APD would just have to wait until JWST was ready to launch before getting a serious start on building WFIRST. Given that it was estimated that WFIRST would take about 7 years to develop, build, and test, this meant that the earliest WFIRST could fly would be about 2025, halfway into the next decade.

At the NAC Science Committee meeting on April 21, 2011, we learned from Rick Howard, the JWST replan lead, that if the JWST budget was limited to $375 million a year from FY 2012 on out, then the 2018 launch date would not be met after all: $375 million a year would push the launch date out to "twenty-twenty-something," endangering the value of flying the mission. So where would the extra funds needed to launch JWST by Bolden's 2018 date come from? The APD could expect to suffer the most, but other SMD Divisions might share the pain as well. The Planetary Science Division had already noted that it could not afford the top-priority flagship mission in the 2011 Planetary Science Decadal Survey, which called for the Mars Astrobiology Explorer-Cacher (MAX-C), costing as much as $3.5 billion. PSD had only $1 billion or so to spend, so there was no sense in asking PSD for spare change.

Ed Weiler ominously declared that some missions that were already largely built and nearing launch might have to be canceled in order to address the budgetary shortfall in SMD. It might even come down to competing one Division's plans for their next flagship mission against the plans for another Division's flagship, not a pretty prospect. ESA had a Cosmic Visions process for competing astrophysics missions against planetary science and heliophysics missions, but NASA did not: instead, it had three separate NAS Decadal Surveys to cover these three areas. Did we need a Decadal Survey of Decadal Surveys? What a battle that would be—there would likely be vicious food fights at the Beckman and Keck Center cafeterias.

The Envelope, Please . . . And the Winner Is: The "Exploring Strange New Worlds" meeting on exoplanets was held in Flagstaff, Arizona, on May 2–6, 2011. On the first day, Wes Traub, a JPL Exoplanet Exploration Program scientist, told me that when he gave his talk on the fourth day he would present an estimate of the frequency of Earth-like planets around solar-type stars, η_E (pronounced "Eta Sub Earth"), the number that was the prime reason for Kepler's existence. While the Kepler team was still busily collecting and processing data and did not wish to try to guess the answer this early in the mission, Wes was willing to extrapolate, on the basis of the miniscule amount of Kepler data made public to date, and try to guess just what η_E was. He came up with the astonishingly high guess of 48% for η_E.

This was astonishing on two counts: first, because it meant that half of the stars in the sky might well have living planets, but also because two of his colleagues at JPL, including Mike Shao of SIM fame, had made their own estimate recently and had come up with a value of η_E of 2%. Their two estimates varied by a factor of 24, quite large, even by astronomical standards, where factor-of-two differences in values are commonly ignored and declared to be the same. These two estimates, though, appeared to put the best bounds known to date on η_E: somewhere between 2% and 48% of sun-like stars might well have Earth-like planets.

Even the lower value would be a complete validation of the Kepler Mission, but the higher value was enough to make one absolutely dizzy when gazing at the night sky. *Essentially all those points of light have planets that might hold life.* How could one look at the night sky again without thinking about this incredible fact? And who was out there, looking back at us?

Cambridge University theoretical physicist Stephen Hawking had recently warned that we should not be sending out messages to the universe, broadcasting our existence; malevolent aliens might decide to pay us an unwanted visit as a result. While some Russian astronomers had indeed concocted and broadcast radio-wave Messages to Extraterrestrial Intelligence (METI), the powerful military radars used to search the skies for incoming intercontinental missiles had already done an excellent job of broadcasting our existence and location to the Milky Way Galaxy and its inhabitants. Somewhere an extraterrestrial SETI search had probably already heard our military radar signals and were puzzling over their meaning: were these caused by intelligent beings, or where they some weird new astrophysical phenomenon?

Most SETI researchers were content to listen for other civilizations. The Allen Telescope Array (ATA) in Hat Creek, California, had grown to a total of 42, 6-m, dishes, thanks in part to a $25 million donation from Microsoft's Paul Allen. However, the SETI Institute folks had spent all of that and another $25 million on the ATA and were now starved for operating funds, much less the cost of building out the ATA to the anticipated final array of 350 radio-wave antenna dishes. SETI was anxious to use the ATA to listen to the most promising exoplanets being discovered by Kepler and had started a program to raise $5 million to do just that. The ATA was able to go back into operation later in 2011, thanks to public donations and support from the U.S. Air Force, which wanted to test the use of the ATA for their own purposes.

Meanwhile, the radio astronomers in Green Bank had decided to return to their SETI roots and use the brand-new giant Robert C. Byrd Green Bank telescope, the largest steerable radio telescope in the world, to search for signals from 86 of the most promising Kepler exoplanet candidates. The race mentality had spread to SETI, thanks to the earmarks for science spending by West Virginia's Senator Robert C. Byrd, the longest-serving U.S. Senator and frequent Chair of the Senate Appropriations Committee, who had passed away the year before.

Lowest Common Denominator: The STScI Newsletter arrived in the mail on May 11, 2011, with the lead story by the STScI Director dealing with the JWST debacle. As the STScI would run JWST, the Director had an obvious incentive to paint as attractive a picture as possible of the wondrous science that JWST would do, regardless of the final price tag or of what would happen to the other missions that would be delayed (e.g., WFIRST), or not occur at all (e.g., SIM, TPF-C), because of JWST's higher priority in the 2001 Decadal Survey and considerably higher costs. The Director mentioned in passing that the U.S. astronomical community had received over $130 million in grants over a 3-year period from two NASA space telescopes, Spitzer and Hubble, and hinted that similar financial support would flow from JWST. He closed by saying "I don't see an astronomers' hurricane, leaving devastation in its wake." The waters that had already flooded other NASA projects had not reached the STScI in Baltimore, and with Senator Mikulski's help, they never would.

Rumors began to circulate that if the annual funding for JWST was not increased substantially above $375 million, JWST would not be ready for launch until perhaps 2022 to 2024. Even Kepler's budget was being threatened by the JWST cost overruns. When its extended mission came up for consideration by the NASA Senior Review, the expectation was that Kepler's budget request might be cut almost in half. The flood waters would not recede until JWST was successfully placed in orbit and began operations.

The *Washington Post* published a Letter to the Editor on June 3, 2011, that I wrote in response to an earlier *Post* story stating that the Pentagon had spent more than $32 billion between 1995 and 2009 on weapons systems that were later canceled. My letter noted that NASA had to cancel several planned space telescopes in the same time interval, simply because they could not be afforded. I made the point that we now knew that Earth-like planets were common in the universe, and made the Swiftian modest proposal that perhaps the Pentagon should extend its purview to protecting us from future alien attacks, thereby addressing Professor Stephen Hawking's concerns. If so, I concluded that NASA had some ideas about where and how to find nearby planets with life and invited the Pentagon to get in touch with me. The *Post* whimsically titled my letter as "If the Pentagon aimed higher." I am still waiting to hear from the Defense Department.

6

The President Proposes, Congress Disposes

Progress was all right; it only went on too long.

—James Thurber, 1894–1961

The JWST replan effort was completed by July 1, 2011, and reviewed by Charlie Bolden and SMD. On July 6, a press release from the House of Representatives CJS Appropriations Subcommittee, chaired by Virginia's Frank Wolf, announced that they would be discussing a plan to "terminate funding" for JWST to remove $431 million from SMD's budget for FY 2012. Not only did this Subcommittee want to kill JWST; they evidently were planning on punishing SMD by removing the JWST funding altogether. It would then no longer be a question of waiting for JWST to launch before building the next APD flagship mission: JWST would vanish, along with the funding line that would be needed to fly WFIRST. The astrophysics community leapt into action by sending frantic emails to each other about this looming disaster; my responses were delayed by the fact that I was somewhere over the middle of the Atlantic Ocean at the time, returning from an Origins of Life meeting in Montpellier, France. The AAS and the Association of Universities for Research in Astronomy (AURA) released strongly worded statements condemning this proposed termination and fretting over the wasting of the funds expended to date on JWST, $3.5 billion, should it be canceled, and the loss of U.S. leadership in astronomy—though ESO and ESA might have a few choice words to say about that particular claim.

On July 11, 2011, the House CJS Subcommittee passed the bill that would terminate JWST, with a budget table showing JWST zeroed out and APD at $683 million for FY 2012. The report language went to some trouble to explain that this was being done for NASA's own good, in order to "establish clear consequences for failing to meet budget and schedule expectations." Dropping JWST would alleviate the "enormous pressure that JWST was placing on NASA's ability to pursue other science missions." It was not clear how removing JWST's $431 million from SMD's

FY 2012 budget altogether would relieve any pressure; if that was truly the case, JWST's $431 million should have been left in SMD's budget.

Passage of this bill led to a firestorm of protests from Maryland's Congressional delegation. Senator Mikulski's reaction on the same day led off with the fact that this would "kill 2,000 jobs nationwide." Wait, is that why NASA was building JWST?

Mikulski called for the Administration to "step in and fight" for JWST. Maryland Representative Steny Hoyer responded in a similar vein, calling on his House colleagues to reverse the decision, noting that JWST was responsible for 500 jobs in Maryland, and 250 at GSFC.

Now frantic emails began circulating among members of the Astrophysics Subcommittee given that the APS would be meeting just 2 days later. These emails formally violated the Federal Advisory Committee Act (FACA) rules that governed the APS, requiring that all APS discussions be held during meetings open to the public. As the Chair, I warned the APS membership about those restrictions and pointed out that I was working to expand the agenda for our upcoming meeting to provide more time to learn about and discuss the JWST dilemma. The agenda listed only a half-hour on July 13, 2011, for the APS to hear about the JWST replan effort from Rick Howard, a ridiculously short time period considering what we had just learned about Frank Wolf's modest proposal.

FACA committees tend to discuss important matters thoroughly, deliberately, and at length. A half-hour might not even be sufficient for the APS to decide where to meet for dinner that evening. However, FACA committees are also constrained about their agendas, which must be printed in the Federal Register 30 days before each meeting, so my entreaties to the NASA HQ folks about reworking the agenda fell flat. The JWST replan flow chart to be presented by Rick did not show the APS as being one of the bodies that the replan needed to satisfy, so our opinions would likely be of value only if they were accepted by the NAC Science Committee, and then by the NAC itself: the NAC presumably had the Administrator's ear.

Shortly before the APS meeting started on the morning of July 13, 2011, Ed Weiler called me into his office for a private meeting where he presented his side of the JWST debacle. Ed felt that he and Jon Morse had been "thrown under the bus" by the Casani report, which blamed NASA HQ more than it did Northrop Grumman, the prime contractor, where the problems arose. Ed noted that he had become so infuriated by NGST's constant requests for additional funds that he had proposed a cancellation review to Charlie Bolden in April 2010. The direct result of that threat was the creation of the Casani TAT, followed by the Casani ICRP, and then the JWST replan we would hear about that day. Shortly after we heard Rick Howard's JWST replan presentation that afternoon, we heard a news flash that an amendment offered by California's Representative Adam Schiff to provide $200 million for JWST in the FY 2012 House budget had been rejected on a voice vote. Representative Wolf had argued that JWST might cost as much as $8 billion before calling for the voice vote, cryptically adding that "We want to do it, but we want to do it in the right way." Was his threat to cancel JWST merely a way to establish a strong negotiating position before meeting with his Senate counterpart, Barbara Mikulski, on a compromise bill? If so, he had certainly gotten everyone's attention.

Howard's JWST update to the APS had specified that they were working toward an October 2018 launch date but did not give an estimate for the total cost to launch, other than that it would be more than $6.5 billion. Our APS Letter Report thanked Rick for his efforts while requesting a lengthier briefing about the replan at our next meeting: the APS intended to keep a close eye on JWST, although recognizing that we had little or no real influence over the fate of JWST. There was really nothing that we or the NAC Science Committee could do except offer support for the science that JWST would accomplish and wait to learn more about what would happen to JWST.

One to Four Habitable Super-Earths: A news article in *Nature* on July 19, 2011, summarized the dim prospects for JWST and WFIRST that emerged at the APS meeting but noted that JWST would be able to "spot Earth-like planets." Well, no, it would not. When the reporter forwarded me a link to his story, I replied that JWST was quite incapable of doing what he claimed it could and that at best, JWST would be able to do what Spitzer had been able to do for transiting planets. As Drake Deming had said at the UCSB KITP meeting the previous year, the estimate was that TESS would find about eight nearby, transiting super-Earths, and that JWST would be able to detect two biomarkers, water and carbon dioxide, in one to four of them. A European group published in 2011 an estimate that JWST might be able to detect another biomarker, ozone, in perhaps one transiting warm super-Earth, at the cost of 2% of the total JWST observing time in the first 5 years in space. Both estimates assumed that TESS would be approved, fly first, and find the targets for JWST. If a star shade could be flown along with JWST, that might enable JWST to image Earth-like planets; but major technical difficulties would have to be overcome, along with a huge increase in cost, which was simply not in the deck of cards that NASA had available.

That was pretty much what the exoplanet community could expect to gain from JWST. As a member of the science organizing committee (SOC) for a conference entitled "Frontier Science Opportunities with the James Webb Space Telescope," I had noticed that there had been not a single abstract contributed to this JWST conference on the topics of exoplanets, star formation, or planetary science. Clearly, other than the invited speakers on these topics, there was not much interest in those three communities about what JWST would do for them. Soon after the Casani ICRP report was released in November 2010, the "Frontier" SOC was told that the decision had been made not to hold the meeting at the Jackson Lake Lodge, just north of Jackson Hole, Wyoming, the lair of idle billionaires and Federal Reserve Board meetings. A "lower profile, less costly venue" was needed, and that meant the meeting would be held at the STScI in Baltimore, Maryland.

On July 26, 2011, NASA HQ announced that Jon Morse would be leaving NASA to accept a position at Rensselaer Polytechnic Institute in upstate New York. There were quiet cries of jubilation from some Astro 2010 folks, who blamed Morse for the pickle that the APD was in. JWST managers at GSFC had already been reassigned as a result of the replan purge, and now the purge had reached NASA HQ. But how much higher would it go?

Charlie Bolden confided in a private meeting with members of the NAC Science Committee at NASA Ames on August 2, 2011, that NASA now had three overall goals: completing the International Space Station, developing a new heavy-lift launch vehicle, and JWST. While that was the order in which he listed them, he took pains to note that they were not listed in priority order. But from where would the additional funding needed to finish JWST by October 2018 be taken? That was still unclear.

At that same NAC SC meeting, Paul Hertz, then the SMD Chief Scientist, presented his take on the lessons learned from Astro 2010. One important lesson was that Astro 2010 had recommended several flagship-class missions and but no medium-class missions, which were about the only missions that NASA could afford to fly these days. This was a lesson that Astro 2020 would need to take seriously: think about probe-class missions, less than $1 billion. Astro 2010 also recommended a specific flagship mission, WFIRST, as the top priority, rather than specifying the top-priority science and letting NASA HQ figure out how best to accomplish that science, as had been the previous approach taken by Decadal Surveys. Perhaps worst of all, Astro 2010 had based their priorities on a Decadal budget that was higher than that finally suggested by NASA, and it was not able to adjust to the JWST cost and schedule problems that were emerging simultaneously with the appearance of the Survey report. Astro 2010, titled "New Worlds, New Horizons," offered no advice about how to proceed in this brave new world.

Shortly thereafter, on August 16, 2011, NASA announced the stunning results of the JWST replanning effort: JWST would require a total expenditure of $8.7 billion, 8.7 times the buy-in cost back in 2000. A total of $3.5 billion had been spent so far, and another $5.2 billion would be needed by the new launch date in 2018. That worked out to about $2 billion to $8 billion per super-Earth that JWST could hope to study, assuming that TESS was selected and performed successfully. Where would the $5.2 billion come from? The rumor was that NASA would make Astrophysics pay half the added cost and spread the rest of the pain to other branches of the agency.

At least one astronomer went rogue and suggested that JWST should be canceled. This was met by orders from Senator Mikulski's staff to "shut up." Mikulski was planning on having the human space flight side of NASA help pay for JWST. *Science* quoted me in my role as APS Chair as saying that the need for $5.3 billion beyond the $3.5 billion spent to date might make it easier for Congress to halt JWST, as it "weakens the argument that NASA has too much already invested in JWST to drop it." Would I be hearing from Mikulski's staff too?

Bolden's elevation of JWST to one of NASA's top three priorities meant that JWST had become "too big to fail," in the words of the infamous phrase used to argue for federal bailout funds for several of the mega-banks whose questionable actions had led to the Great Recession.

The Sun Is Not a Typical Star: In its hunt for exoplanets, the Kepler Mission was also revolutionizing our understanding of stellar structure and evolution. Kepler was providing measurements of the brightness changes of a host of different stellar types, measurements that were as detailed as those of our own star, the Sun. The unfortunate outcome of this new information was the realization that our Sun is not a typical solar-type star. The Sun is considerably quieter than its colleagues, which seemed to be quite a bit noisier, probably because of having more star spots on their surfaces.

Kepler had been designed on the assumption that the Sun was a typical solar-type star. As the solar-type star for which we had the most information prior to Kepler, there really was no other assumption that could be made. The unanticipated discovery that stars were noisier than expected meant that Kepler would need to work for at least 8 years, instead of the 3.5 years in the prime mission plan, in order to take enough transit data to beat down the noise and detect Earth-like exoplanets.

At a Kepler Science Team meeting on August 18, 2011, at the SETI Institute close to NASA Ames, we learned that a dozen Kepler team members had been let go already because of a decrease in the FY 2012 budget. Kepler team members worried about getting the extra funds needed to keep Kepler working in a NASA that put JWST ahead of every other science mission in the entire Agency. The 2012 Senior Review, where Kepler would compete against other NASA space telescopes, like HST, for future funding would be critical, not only for keeping the Kepler team at work but for getting the best possible estimate of η_E. Much of the meeting was devoted to a proprietary discussion of Kepler's funding dilemma: Kepler's nominal FY 2014 budget was $0.056 million, about enough to empty the project offices and turn out the lights. Only a successful Senior Review outcome could prevent that from happening.

The good news from Kepler was that there were now 1,275 candidate exoplanets, 54 orbiting in their HZ, and 68 of Earth size. There was even a hint of the first HZ planet, Kepler Object of Interest KOI-87.01, eventually to be known as Kepler-22b. Bill Borucki was busily writing a paper about this major Kepler discovery.

Ground-based telescopes continued their successful assaults. Michel Mayor's HARPS team announced on September 12, 2011, the discovery of 50 new planets, including 16 new super-Earths, and estimated that about 40% of solar-type stars have such worlds. Francesco Pepe and the HARPS team had also found what was perhaps the second super-Earth in the HZ of a red dwarf star, HD85512b, with a mass at least 3.5 times that of the Earth. The press release crowed that ESO's HARPS was the "world-leading planet hunter," having discovered more than 150 exoplanets, including about two-thirds of those with masses less than Neptune, about 17 Earth masses. Ground-based astronomers were not about to wait for space to catch up, but Kepler's 1,275 exoplanet candidates would soon swamp the Doppler survey numbers.

Who Wants to Pay for Something They Did Not Order?: The rumor was that OMB insisted that JWST's cost overruns be borne only by NASA's four science divisions. Planetary scientists began to call for an open debate about whether JWST was too big to be canceled. It was clear that in an era of flat NASA budgets, the additional funds required for the replanned JWST could only come from within NASA, and the planetary scientists realized that their future mission plans were in the same jeopardy as those in astrophysics. An editorial in the September 8, 2011, issue of the *Planetary Exploration Newsletter*, signed by over a dozen prominent figures, stated that they "reject the premise that JWST must be restored at all costs," and called for an assessment across all four SMD divisions of the science that would be lost to keep JWST alive.

The JWST aficionados immediately returned fire on September 11, 2011, with plans for a Town Hall Webinar that would "clarify the costs of the mission" and "discuss the science enabled by JWST." The STScI Director trotted out a reworked version of the same article written for the STScI Newsletter earlier that year, once again invoking meteorological phenomena and repeating that "I don't see an astronomers' hurricane, leaving devastation in its wake." To whom could he be referring? Beats me. The AAS Division for Planetary Sciences returned fire with a statement that called for consideration of the "scientific bounty reaped by planetary missions." The AAS itself then decided to back off of its previous call for support of JWST and stated that the "AAS does not support any one Division or astronomical discipline above others, or to the detriment of others." Instead, the AAS made the bold statement of their support for "lobbying for generous support of all of astronomy."

A few days later, on September 14, 2011, Senator Mikulski made sure that JWST would be fully funded in the Senate appropriations bill for 2012, even if it meant curtains for other NASA science missions. Mikulski wrote in $530 million for JWST, $156 million more than was requested, out of a total NASA budget of $17.9 billion, which was $509 million less than in FY 2011. But where would that $509 million come from in a reduced NASA budget? The Senate bill did not specify what should be cut—Charlie Bolden might be handed that unpleasant task, if this markup held in the conference committee.

The House, however, continued the battle over funding JWST at the expense of other NASA activities. Frank Wolf sent a letter to the OMB on September 28, 2011, asking what programs they proposed to cut in order to pay for JWST: if there was no response from OMB, Wolf would conclude that JWST was not a higher priority than other NASA programs. Wolf was not willing to take all the heat for dropping

JWST. The House-Senate conference committee would meet in a few weeks and settle the issue, one way or the other.

The day before, September 27, 2011, SMD Associate Administrator Ed Weiler announced his plans to retire and follow Jon Morse in leaving NASA. Weiler had spent 33 years at NASA, including nearly 20 years as the Chief Scientist of Hubble. I had always been impressed in my dealings with Ed through the years: he was someone who knew the agency like no one else. One close colleague of mine commented privately that "Ed was served up like John the Baptist as the price for Webb staying in the NASA budget." Although Ed had been planning his retirement for some time, the word was that he had hoped to stay on at HQ long enough to witness the successful landing of the Mars Science Laboratory (MSL) in August 2012, a planetary flagship mission that he had shepherded to the launch pad. MSL was safely on its way to Mars, but Ed would soon be on his way south on Interstate 95 to retirement on the east coast of Florida, well south of Cape Canaveral and the Kennedy Space Center.

Before his own departure from NASA HQ, Jon Morse had asked me to stay on for a 3rd year as the Chair of the NAC Astrophysics Subcommittee, and I had agreed to do so, thinking that the Astrophysics Division needed as much continuity as possible in these contentious times. Ed's decision to retire reinforced my intention to continue to try to help "advise" NASA about its predicament.

Jumping the Dark Energy Gun: Three of the world's leading dark energy enthusiasts were announced to be the winners of the 2011 Nobel Prize in Physics for their leadership of the ground-based studies that had led to the dark energy concept. Two of these three, along with their team members, who had shared in the Nobel glory, were perhaps expecting to ride WFIRST all the way to Stockholm, but their ride arrived over a decade earlier than expected. Now that the dark energy Nobel Prize had already been announced, some wondered if that meant that we did not have to fly WFIRST anymore. The field of exoplanets, with arguably sounder, more dramatic discoveries, regarding life in the universe under its expanding waistline belt, was still waiting for a Nobel Prize, or a Crafoord Prize, or a Kavli Prize, or any sort of prize from a wealthy Scandinavian.

On the same day as the Physics Nobel Prize winners were announced, October 4, 2011, ESA announced that Euclid had been selected as one of its next two medium-class missions. Euclid would be a 1.2-m telescope that would launch in 2019, well before WFIRST could even hope to get started, given a late 2018 launch for JWST. Apparently Europe was drunk on dark energy; planet-hunting PLATO would have to wait for another chance in the coming years. Not so in Washington: NASA HQ began to speak publicly about the fact that WFIRST was in mortal danger because of the dire budget situation, with only JWST being assured of sufficient funding. Flying JWST might mean cutting back on nearly everything else in the SMD at NASA. Rick Howard presented a "breach report" to Congress on October 24, 2011, required by JWST's cost growth of over 30%: since 2009, JWST costs had in fact grown by 140%. The breach report did not state which NASA missions would suffer delays or cancellations as a result of JWST's replan.

Maryland Representative Donna Edwards noted that the OMB seemed to be deciding what to cut in NASA's portfolio, and she argued that plans for a joint NASA-ESA Mars exploration program were in jeopardy, oddly resisting the precedent set by other Maryland Senators and Representatives to support JWST, no matter what.

Point, Mikulski: On November 15, 2011, the House-Senate conferees filed their report on FY 2012. NASA would lose $648 million overall, but SMD would gain $155 million, all of which would go to fund JWST at $530 million, the number called for by Senator Mikulski's Senate markup. Funding for space exploration and space operations would decrease, and the Space Shuttle fleet would be retired. Someone had to pay for JWST, and the human space flight side of the house would be helping out. SMD would also help pay for JWST, with significant cuts allocated to Planetary and Earth Science, as well as Astrophysics.

A few days later, on November 21, 2011, Charlie Bolden picked astrophysicist and former astronaut John Grunsfeld to succeed Ed Weiler as chief of the SMD. Weiler gave Grunsfeld his stamp of approval. Grunsfeld, formerly the deputy director at STScI, could be expected to be just as much of a JWST-hugger as he had been a Hubble-hugger, with the difference being that JWST's orbit at an Earth–Sun Lagrange point meant that NASA astronauts would not be able to service JWST, or to repair it, in case of a severe vision problem like that with HST. Given Grunsfeld's background in astrophysics, there was some concern that NASA's three other science divisions might not get a fair shake; but Weiler said that as long as decisions were made on the basis of peer review, there was no need to worry—Ed had faced the same concerns when he came the head of SMD.

Kepler Strikes Again and Again and Again . . .: A press conference was scheduled on December 5, 2011, for Kepler's latest discovery, this time at NASA Ames rather than at NASA HQ. Bill Borucki's discovery of the smallest-radius exoplanet found to date in the HZ of another star, Kepler-22b, would be the main feature. Kepler-22b has a radius about 2.4 times that of the Earth but had an uncertain mass. The target star, Kepler-22, however, was a G dwarf star similar to the Sun, making Kepler-22b the first HZ planet, possibly a super-Earth, found in orbit around a solar-type star, unlike the previous HZ super-Earths found by ground-based Doppler searches on red dwarf stars. The Kepler press conference signaled the beginning of the First Kepler Science Conference at Ames, where the stunning results garnered to date would fill a 5-day-long meeting. Kepler had now discovered over 2,300 candidate exoplanets, in just the first 16 months of data. Many more remained to be found, patiently sitting buried in the many terabytes of digital data that was spinning endlessly on the disk drives at the Kepler Science Center at NASA Ames.

Kepler had now confirmed 27 transiting exoplanets, and the list could only grow. With these confirmations, it was becoming clear that Kepler had found an exoplanet oasis in the middle of the discovery space desert predicted by theoretical models of planet formation in the core accretion scenario. As noted at the Torino meeting in 2010, Doppler surveys had already found evidence of the oasis; and Kepler was also finding an abundance of short-period super-Earths, objects that should not be commonplace, if the conventional wisdom about planetary system formation was correct. Evidently it was not.

7

Open for Business, Under New Management

I spend money on war because it is necessary, but to spend it on science, that is pleasant to me.

—King George III, 1738–1820

John Grunsfeld took over as head of NASA's SMD on January 4, 2012. A few weeks later, on January 24, I gave an invited talk about my research to the Astrophysics Division astronomers at NASA's GSFC, and afterward I had the honor of a personal tour of the JWST assembly areas given by Goddard's John Mather, a 2006 Nobel Prize winner for his work on mapping the cosmic microwave background radiation field, and, incidentally, the Project Scientist for JWST. JWST was coming along fine, though all I could see at that time was the backend of the telescope, where the science instruments were being mounted, prior to the installation of the 18 hexagonal mirrors.

When I returned to DTM, I found an email from the new acting head of the Astrophysics Division at NASA asking for me to call him. I dialed his cell phone number and reached him as he was pumping gas into his car on the drive home. He quickly cut to the chase and let me know that Grunsfeld had decided to "clean house" at HQ, and that as a result, I would no longer be the Chair of the NAC Astrophysics Subcommittee after March 2012 instead of serving for a 3rd year, as I had agreed to do when asked by Jon Morse. That also meant I would no longer serve on the NAC Science Committee.

Ohhhh-kayyyy. First Jon Morse, then Ed Weiler, and now lowly Alan Boss? I found the news to be considerably more amusing than disappointing. Chairing the APS was a lot of work—the chair does the real work on most committees—and I was not even being reimbursed by HQ for my lunch costs (at DTM, I bring my lunch), or my parking expenses in the garage under HQ, during the days I spent there working literally for coffee and cookies. I had a great day job as a theoretical astrophysicist at DTM, as well as a night job as an observational astronomer, and I had been neglecting somewhat those duties to work on "advising" NASA. Sayonara, baby.

Wes Huntress, the Chair of the NAC Science Committee, was surprised and unhappy to hear about my summary dismissal from his committee, as his understanding was that any staffing changes were to be made in consultation with him. He asked me to continue to chair APS meetings and attend Science Committee meetings through April, and I agreed.

We learned in early February 2011 that the President's FY 2013 budget request for NASA called for flat funding overall, but an increase to $628 million for JWST, coupled with a $300 million decrease for Planetary Science, a cut large enough to kill any hopes for a joint NASA-ESA Mars exploration program. Instead of the MAX-C mission, the highest-ranked large mission in the 2011 Planetary Sciences Decadal Survey report, the President's budget called for a sample return mission to an asteroid. An asteroid? The budget request noted that WFIRST and MAX-C were "deferred" as they were "unaffordable." The top-priority missions of *both* the Astrophysics and Planetary Sciences Decadal Surveys had thus been "deferred" as a result of the flat overall budget and the higher agency priority of Hurricane JWST.

A New Sheriff Arrives in Town: John Grunsfeld took a more relaxed approach to running NASA's Science Mission Directorate than Ed Weiler did, wearing polo shirts at HQ meetings, rather than a dress shirt and tie, and presenting more as an astronaut than as an administrator. He opened his first meeting on February 24, 2012, with the NAC Astrophysics Subcommittee with his admonition that as Special Government Employees (SGE) on the days that we were serving on the APS, we had "lost the right of free speech" and "APS members should watch what they say." So much for giving NASA "advice." This would be my last meeting as APS Chair, so I was happy to sit back and marvel at his remarks. The next chair of the APS would turn out to be a personal friend of Grunsfeld and a member of the JWST advisory committee, presumably someone who would have the proper viewpoint on the overriding importance of JWST for all of NASA's science efforts. The inside word was that I did not have "the right stuff" regarding JWST.

Grunsfeld repeated his advice about watching what you say at his first NAC Science Committee meeting on March 6, 2012, though this time there were other free thinkers in the room who were willing to give NASA advice that might not be desired. Scott Hubbard, for one, had bemoaned the loss of the joint NASA-ESA Mars program, telling the *Washington Post* that it was "a scientific tragedy and a national embarrassment."

Charlie Bolden attended the meeting as well and pointed out that he wanted the Science Committee to be critical of NASA's plans but helpful. What could be more helpful than indirectly alerting the President and the White House to a major, looming problem in NASA's Astrophysics Division, a problem that turned out to be even worse than first thought, a problem that led the Administration to throw enough precious resources at it to make sure that JWST would survive?

Personally, I was beginning to believe that my Hurricane Katrina remark had raised the visibility of the JWST specter high enough to help save it from outright cancellation.

Science magazine mistakenly reported on March 16, 2012, that JWST's price tag had increased to $18.7 billion instead of the correct cost of $8.7 billion. Apparently the $18.7 billion typo did not seem to be any more outrageous than $8.7 billion, and slipped by *Science*'s proofreaders. On March 28, Bolden told Senator Mikulski's Budget Subcommittee that if the federal sequester being threatened for January 2013 took place, NASA would have "to put all our funds on the priorities and forget everything else." As one of the Agency's top three priorities, JWST would survive, but that might be it for any other space science missions.

Bill Borucki sent an email to the team on April 2, 2012, with the good news that Kepler's proposal to the 2012 Senior Review had been given the green light for an extended mission. Bill concluded the email with "Hooray for Kepler!," which signaled joyous delirium by Bill's standards. Provided that Kepler did well again in the next Senior Review in 2014, Kepler would stay alive for 4 more years, for a total of 7.5 years, as needed to get a good estimate of η_E, given the noisy stellar hosts.

The Senior Review, though, did say that Kepler would have to cut its costs significantly, and that meant fewer funds for the scientists working on Kepler data. The budget was so tight that there would not even be funds to help support the Northern Hemisphere (N) clone of the HARPS Doppler spectrometer. HARPS-N was planned to begin observations in 2012 on the 3.6-m Italian National Telescope "Galileo" on La Palma in the Canary Islands. HARPS-N was intended in large part to determine the masses for Kepler planet candidates; the original HARPS, now called HARPS-S, was in Chile, too far south to observe Kepler's targets in the constellations of Lyra and Cygnus.

A Senator by Any Other Name: On April 5, 2012, the STScI renamed its "Multi-mission Archive for Space Telescopes" to the "Barbara A. Mikulski Archive for Space Telescopes," preserving the existing acronym MAST. Senator Mikulski humbly accepted the honor, intended to "establish the Senator's permanent legacy to science." Curiously, the previous name for MAST appeared to have been thoroughly scrubbed from the Internet: a Google search for the exact phrase "Multi-mission Archive for Space Telescopes" turned out an amazingly low two Web entries. Apparently the history of MAST had been successfully rewritten. The renaming was a small token of appreciation for Mikulski's huge success at keeping the STScI in business for decades to come.

The Swiss-European Doppler planet search team announced the results of their searches for HZ super-Earths around red dwarf stars: η_E is about 41% for these low-mass stars, a number similar to that being tossed around for the mostly solar-type stars being studied by Kepler. The March 28, 2012, ESO press release headline started off with "Many Billions of Rocky Planets in the Habitable Zone." Red dwarf stars make up the great majority, about 80%, of the 100-billion-odd stars in the Galaxy. Doing the math, the ESO press release was thus claiming the existence of about 32 billion HZ super-Earths in the Milky Way, give or take a few.

The TCU prediction of a crowded universe was being verified by both the Doppler spectrometers on the ground and by the results emerging from the Kepler telescope in space. Kepler was slowly moving farther and farther away from Earth, as it orbited the Sun on its own, far from the grasp of Earth's gravity, but dutifully responding to the commands sent to it from the spacecraft control room in Boulder, Colorado, close to where it had been built by Ball Aerospace. We learned at a team meeting at Ames on April 19, 2012, that the Sun was causing problems for Kepler again: this time it was a flurry of solar coronal mass ejections leading to energetic particles striking the Kepler detectors a few days later, once the solar wind reached Kepler's orbit, and causing an increase in the background noise. First the Sun was deceptively quiet, compared to other solar-type stars, and now the Sun was raising hell with Kepler's transit data.

Out of the Blue: NASA revealed on June 4, 2012, that it had been offered two complete telescope optical systems, originally built for spy satellites intended for the National Reconnaissance Office (NRO) but never used because of the skyrocketing costs of the canceled top-secret program, called the "Future Imagery Architecture." The telescopes were wide-field versions of Hubble: each one was 2.4 m in diameter, with optical specifications that could not be fully revealed, even to NASA, because of the NRO's secrecy rules. Such a telescope would be perfect for the WFIRST mission, which had been considering 1.1-m and 1.3-m primary mirrors rather than the 1.5-m primary mirrors specified by Astro 2010. A mirror twice the size would give WFIRST four times the light-gathering power, speeding the rate of discovery and possibly allowing the addition of new science goals. Adding a coronagraph, for example, would enable WFIRST to snap photos of large exoplanets around nearby stars.

NASA had been quietly considering the possible uses of the NRO telescopes ever since they were offered confidentially to NASA in January 2011. After 1.5 years of in-house deliberation, NASA HQ went public. Astronomers were stunned by the manna from heaven that rained from the skies. There was one firm stipulation about NRO's offer, however: *the telescopes could not be used to look at the Earth*, their original purpose. NASA's astrophysicists had no problem with that ground rule: no problem at all.

Each of the optical assemblies was estimated to be worth about $250 million, and they had been sitting unused in a clean room in northern New York state for the last 10 years.

While WFIRST was still on hold, perhaps these unexpected gifts would provide the leverage needed to get WFIRST back on track. They might function well as a "probe-class" exoplanet imager, not quite as powerful as the plans for a 3.5-m by 8-m elliptical TPF-C primary mirror, but an important step in the direction of a direct-imaging space telescope for nearby exoplanets. With two telescopes available, the intoxicating vision arose of using both of them, one for WFIRST's Astro 2010 goals and one for exoplanet imaging. One early estimate by NASA HQ was that a WFIRST using one of the NRO 2.4-m assemblies would cost between $1 billion and $2 billion and might be ready to launch in 2020.

The previous week, I had attended meetings of the ExEP Technology Assessment Committee (TAC) at Ames and JPL, where we reviewed laboratory work being done on developing improved coronagraphs and star shades for exoplanet imaging. I sadly noticed that JPL's Building 301, where we had often met in the 1990s and 2000s, still had large posters hanging on the walls showing SIM-Classic, TPF-C, and TPF-I, all three now long gone, but evidently not quite forgotten.

A Green Light Means Go: ESO announced on June 12, 2012, that the E-ELT had been approved by its governing council. The E-ELT had been downsized slightly, from a diameter of 42 to 39 m, and its cost was now estimated at $1.35 billion. The E-ELT would be built on Cerro Armazones, close to the existing Paranal Observatory in northern Chile; and the plan was that by the early 2020s, the E-ELT would begin searching for "habitable alien planets." Searching for alien worlds was considered to be a major selling point, along with dark matter and dark energy.

ESO needed to make sure that it had at least 90% of the funding it needed from its member countries in order to begin construction. Europe was in a deeper Great Recession than the United States was, and so it was amazing that ESO was still gung-ho about the E-ELT in a budgetary environment where the unfamiliar phrase "sovereign debt" had become an international buzzword.

Kepler Loses a Reaction Wheel: We heard some bad news on a Kepler science team telecon on July 19, 2012, led by Natalie Batalha, effectively the new science team leader. One of the spacecraft's four reaction wheels had failed on Friday, July 13. At first the thought was that the recent solar storms had created the anomaly in the data, but it turned out to be the loss of reaction wheel #2. It was well known that once a reaction wheel failed, it would not recover, though the Kepler engineers would try anyway. They soon decided that the ball bearing cage on #2 had broken: #2 was now long gone.

Three other wheels were still working: #1, #3, and #4. All three were needed to ensure the steady pointing that Kepler required in order to achieve the precision needed to detect transits by Earth-size planets in front of solar-size stars, that is, photometric precision much better than 0.01%. It was learned just prior to launch in 2009 that these reaction wheels were not as reliable as desired, as the same design had failed on another NASA mission; but there was no time left to replace them, only to make some minor changes in the hope of extending their lifetimes on orbit. The engineers were now trying to stabilize Kepler's pointing with the three good wheels; and once they were able to do so, and start taking science data again, we would learn what the effect would be on Kepler's key measure of instrument performance: photometric precision.

From here on out, the Kepler team would have to watch the remaining three reaction wheels, looking for any signs of failure. The engineers would be making some changes in the way that the reaction wheels would be treated, largely in order to try to coax the lubricant to more properly coat the moving parts. Reaction wheel #1 had already made some 1.5 billion revolutions and was approaching the dreaded "mean time before failure" measured in revolutions. Once another wheel died, Kepler's prime and extended missions would be effectively over, no matter what the next Senior Review said.

A few months later, Bill Borucki sent a poignant email to the Science Team mourning the passing on September 12 of Dave Koch, Bill's partner in morphing the 1992 FRESIP dream into the reality of the Kepler Mission. Koch had been suffering from ALS for years but continued to work on Kepler until the end.

Should We Write the NRO a Thank-You Letter?: The question of what to do with the NRO 2.4-m telescope optics was the subject of a meeting at Princeton University called by Princeton's David Spergel. As a strong advocate of both dark energy and exoplanet searches, Spergel was a natural leader for bringing the community together to discuss what would be the best use of the NRO gifts. Meeting in Peyton Hall on campus on September 4–6, 2012, the main players presented a wide array of ideas. David led off by proposing that WFIRST would cost about $2 billion if it used the NRO optics and had a coronagraph that could serve as a TPF precursor, a bit more than the $1.6 billion estimate for Astro 2010's 1.3-m design for the first WFIRST Design Reference Mission (DRM-1).

Michael Moore from NASA HQ noted that Charlie Bolden had to decide by July 2013 whether to accept the NRO gifts, and if so, how best to use them: the final decision was his. The NRO gifts would be an agency asset, not an APD asset, but a "Dear Colleague" letter had been issued by APD the month before calling for ideas on how best to use them, on an "as is" basis, in order to try to minimize costs. An engineer from ITT Exelis, the firm that built the assemblies, told us about their specifications, including the strange asymmetric, six-arm, spider struts used to support the secondary mirror above the primary. Only one of the optics sets was complete, including thermal hardware, which was necessary because the mirror coatings could only remain stable above about 200 K (Kelvin).

Alan Dressler gave the blessing of the Astro 2010 EOS panel for the NRO option for WFIRST, which would clearly enable more science than the 1.3-m DRM-1 design. Princeton's Jeremy Kasdin noted that progress had been made in designing coronagraphs that would work with obscured optics like that of the NRO assemblies, where the secondary mirror is centered on the primary mirror, held by those six ungainly struts, giving hope that this nonideal design might still be useful for imaging exoplanets, if not exoEarths. Two rapidly deformable mirrors would be needed to suppress the stray star light diffracted by the obscuration, allowing the exoplanet to be seen, which Kasdin estimated would allow extrasolar ice giant planets, cold super-Earths, to be seen.

Colorado's Web Cash, one of the early proponents of star shades, argued that ice giants would not be enough: a star shade would enable WFIRST to find exoEarths as well.

Jon Morse warned that the President's Budget Request for FY 2013 did not show the return of the annual cost of building JWST to the APD budget line in the out years; unless those funds were returned to APD once JWST flew, we would not be able to afford anything in the flagship class, even with the NRO gift.

The first day of the Princeton meeting was a rainy day, with the mid-Atlantic being drenched by the remnants of Hurricane Isaac, which had hit New Orleans, as Katrina had in 2005. I managed to joke about this coincidence with the STScI Director and the JWST astronomer who had not particularly cared for my analogy the first time, but given that JWST had since been given a stay of execution, even they could chuckle a bit now. John Grunsfeld arrived on the last day of the meeting, but I did not get a chance to see if I could get a similar chuckle out of him.

Partying in Reno, Nevada: I was sitting alone in the back of the Tahoe Room at the AAS Division of Planetary Sciences meeting in the Grand Sierra Resort on October 15, 2012, when *Nature*'s Eric Hand asked me if I knew about the discovery of an Earth-mass planet around Alpha Centauri B. Huh? I scurried back to my hotel room to read Eric's email about the embargoed press release from ESO. I had been at the Geneva Observatory a few months earlier for a PhD thesis defense, but my Swiss hosts were tight-mouthed and did not give a hint about this bombshell discovery.

The HARPS-S team had been hard at work in Chile, and the Swiss-European team announced on October 17, 2012, that HARPS had detected an Earth-mass planet in orbit around the Alpha Centauri B star, a member of the triple-star system that is the closest to the Earth, only 4.4 light-years away. The planet was a "hot Earth," orbiting close to its star every 3.3 days, and its mass was estimated to be as low as 1.1 times that of the Earth. If this lower mass estimate was correct, this would be the first detection, not of a super-Earth-mass exoplanet but of an Earth-mass exoplanet.

The closest star system was now known to harbor a likely Earth-mass planet. Humans had stared at the Alpha Centauri system for millennia without any idea that another Earth, albeit so hot that no one would want to live there, was right there in front of them the whole time. What else was out there in the Alpha Centauri system, waiting to be found?

First CoRoT, Next Kepler?: The CoRoT telescope suffered a fatal computer failure on November 2, 2012, a few weeks before Kepler finished its prime mission and began its extended mission phase. In fact, the CoRoT failure occurred just 3 days after the CoRoT mission had been extended for another 3 years. The CoRoT team would try a workaround, but it was most likely dead. CoRoT had found several dozen exoplanets, including CoRoT-7b, and the CoRoT team members were satisfied with what they had been able to achieve.

ESA was interested in flying another low-cost, transit photometry mission, announcing on October 19, 2012, that it planned to launch CHEOPS in 2017. CHEOPS (Characterizing Exoplanets Satellite) would be a small mission, basically the same telescope as CoRoT, with a 12-inch (0.3 m)-diameter mirror, and was intended to search for transiting exoplanets around nearby bright stars that were already known to have exoplanets. Given what Kepler had found, that pretty much meant that CHEOPS should look around *every* nearby star.

Severe funding difficulties were encountered during ESA's budgetary negotiations on November 20–21, 2012. ESA might have to delay a large mission, or cancel a mission extension or a small mission, according to the University of Bern's Willy Benz, chair of ESA's Space Science Advisory Committee. Given that Willy was the Principal Investigator for CHEOPS, and an old friend of mine, I hoped that the funds saved by the early demise of CoRoT would allow CHEOPS to survive.

The extended mission for Kepler began on November 14, 2012, possibly running for another 4 years. However, the Kepler team continued to keep a close eye on their last three reaction wheels, one of which was starting to act up in the same way that the dead wheel had acted before it gave up the ghost. In January 2013, the Kepler team tried the trick of turning them off for 10 days, in the hope that the lubricants would spread throughout the wheel bearings by capillary action. No science data could be taken during the 10-day interval, but that would be a small price to pay if it would bring the reaction wheel back to health.

The Kepler team released another data set to the public, but only after first determining that Kepler had now found 461 more exoplanet candidates, bringing its total score up to 2,740 exoplanets, roughly one hundred times as many as CoRoT had been able to find. CoRoT had enjoyed a lengthy head start in their space race, but Kepler had caught up and long since passed CoRoT in the exoplanet sweepstakes.

Let's Do It Again in 2020: John Grunsfeld announced on December 4, 2012, that NASA would indeed be heading back to Mars, with a launch planned for 2020, skipping the intervening 2018 launch opportunity. The plan was to fly a souped-up version of the Curiosity rover, which had been roaming the Martian surface for the last 4 months. Curiosity 1.0 had cost $2.5 billion; but the second time around, the cost was estimated to be about $1.5 billion. The hope was that Curiosity 2.0 would function as MAX-C had been intended, as a rover that would gather and cache interesting rock samples for retrieval by a later Mars mission. The re-election of President Obama the previous month, and the consequent retention of Charlie Bolden as head of NASA, meant that Bolden's insistence that the long-term goal of NASA should be humans on Mars would have as its first consequence a renewed push for its robotic exploration.

Tau Ceti or Bust: The closest sun-like star to the Sun, Tau Ceti, was announced to have a system of at least five super-Earths, with minimum masses in the range of two to seven Earth masses, on December 19, 2012. Tau Ceti lies just about 12 light-years away and is visible to the naked eye. One of the five was found to orbit in the star's habitable zone, with an orbital period of 168 days. Mikko Tuomi of the University of Hertfordshire was the lead author on the A&A paper, along with Paul Butler and Steve Vogt and a dozen others. Jean Schneider sent out an email announcing the claim, noting that the five were "still to be confirmed," which was fair enough, as Tuomi had had to perform a miracle or two, or maybe five, in order to pull all five exoplanets out of the Doppler data: the title of their paper was "Signals embedded in the radial velocity noise."

The next day, December 20, 2012, the *Washington Post* reported that Senator Mikulski would become the new Chairperson of the Senate Appropriations Committee, as a result of several more senior Senators choosing other chairmanships following the death of the previous Appropriations Chair earlier in the week. Christmas had come early this year to NASA's GSFC and the STScI—they would certainly be having joyous holiday parties.

Just Don't Look Down: In early February 2013, NASA held a competition at the Marshall Space Flight Center in Huntsville, Alabama, to consider a wealth of ideas for using the second NRO optical system. The first 2.4-m system had been assigned to a study team chaired by David Spergel and GSFC's Neil Gehrels for possible use as WFIRST, leaving the second one up for grabs. The JPL ExEP office coordinated a plan to build the best possible exoplanet-dedicated telescope that could be imagined with a 2.4-m telescope, the Exoplanet Observatory (ExO), basically a Hubble-class telescope dedicated to imaging exoplanets. Given the zero cost to NASA of the primary and secondary mirrors, JPL estimated that ExO could be built for a total of $1.5 billion, including a low-cost Falcon 9 rocket launch in 2022 to the same Earth–Sun Lagrange point (L2) where JWST was headed. Many other innovative ideas were considered in Huntsville as well, including an idea to study the Earth's aurorae in ultraviolet light, with the hope that using ultraviolet light to look at the Earth would not violate the NRO rule that the telescope could not look down at the Earth. The proponents had argued that at ultraviolet wavelengths, the Earth's surface could not be seen; but it was unclear if the NRO honchos would buy that argument. ExO looked like a sure winner in comparison, and it turned out to be one of seven concepts chosen for further study on March 4, 2013. ExO would come to life not in a machine shop but in manifold PowerPoint presentations, the birthing routine for all NASA missions. Compared to the $3 billion cost of a single B-2 bomber, ExO seemed cheap; but even $1.5 billion is a lot when you are broke.

TESS Selected for Flight: NASA announced on April 5, 2013, that TESS was chosen for launch in 2017, just ahead of JWST, barely in time for TESS to find a few nearby transiting super-Earths that JWST could then examine for the presence of atmospheric carbon dioxide and water. TESS made sense for NASA: it was low risk and low cost, kept NASA in the exoplanet business, and would find the transiting exoplanets that are the closest to the Earth. TESS would focus on the 1,000 closest red stars, as TESS's planned 2-year lifetime and observational limitations would favor searching for short-period HZ worlds around such stars, not the long-period HZ planets that were Kepler's quarry.

M dwarfs were becoming all the rage. Courtney Dressing and David Charbonneau of the CfA used Kepler data to estimate that 15% of red dwarf stars had HZ Earth-size planets. The same April 10, 2013, issue of the ApJ Letters also contained a paper by Ravi Kopparapu of Penn State, which used a less restrictive estimate of the size of the HZ to estimate a frequency of about 50% for Earth-size planets orbiting M dwarf stars.

A few days later, on April 17, 2013, the Government Accountability Office (GAO) released a study of NASA's performance entitled "Assessments of Selected Large-Scale Projects." The GAO had studied 18 NASA missions that cost over $250 million but deliberately left JWST out of the study, "in part because of its disproportionate effect on the portfolio average." JWST's poor performance would make the rest of NASA look worse than it was, increasing the portfolio's average development cost growth from 3.9% to 46.4%. The GAO had no stomach for including JWST in the key summary graphic, which showed gradual improvements in performance for all the other NASA missions in the interval from 2009 to 2013. The JWST Project responded in a footnote in the GAO report that JWST was staying on track with the milestones set out for the project in the rebaselined plan. It had better—Senator Mikulski's office was still watching.

Close Packing in the HZ: Another Kepler press conference was held on April 18, 2013, again at NASA Ames. The event revealed the most stunning HZ discovery to date: a solar-type star with not one but two super-Earths orbiting in the star's habitable zone. The star, Kepler-62, with a mass 69% that of the Sun, was orbited by at least five transiting planets, two of them within the HZ. Their sizes were just 40% and 60% larger than that of the Earth, making them the smallest transiting HZ planets found by Kepler. Their masses were not well constrained, so these super-Earths could turn out to be largely rocky worlds, worthy of the name "super-Earth," or they might be ice worlds, kept from melting inside by their temperate orbits in their star's HZ. John Grunsfeld called the Kepler spacecraft a "rock star of science," though if these two new worlds should turn out to be low-density ice worlds, would Kepler be an "ice star of science"?

While the Kepler science team was cranking out these fantastic results, the balky reaction wheel #4 onboard the Kepler spacecraft was acting more and more like the failed wheel #2 had been acting before it finally stopped turning. The 10-day rest period had not fixed the problem. The team began to make plans for what Kepler could do if the critical third wheel was lost.

As a result of the federal sequestration that had taken effect, the Second Kepler Science Conference, planned for November 2013 at NASA Ames, was canceled on April 30, 2013. As a co-chair for the meeting, we had already planned a full 5 days of spectacular new results from Kepler and were about to start inviting the lead-off speakers for each session. The news of the cancellation stopped our preparations altogether. The best we could hope for would be to delay the meeting by a year and shoot for holding it at Ames in November 2014. But when would we know if even that was okay with NASA HQ?

Even more bad news soon followed: Kepler's reaction wheel #4 failed on May 14, 2013. A snap press conference was held at NASA HQ on the afternoon of May 15 to announce this sad fact and to put the proper spin on this failure to spin: John Grunsfeld and Bill Borucki cheerfully stated that the Kepler Mission had been a great success and was not over yet. Efforts would be made to restart the failed wheel, but the same efforts had not worked on the identical wheel that had failed first, making the bet of getting a third wheel working again a bet with odds so long that even I would not take it, in spite of my typical horse- and dog-racing strategy of betting on the contender with the longest odds to win. As a former Kepler science team member, I had a dog in this race, but even I would not bet on it.

I had first heard the terrible news just as I was leaving my office to attend a dinner party for the day's DTM seminar speaker. I was poor company during the dinner, as all I could think about was what would a crippled Kepler be able to do now? Would we still be able to get a good estimate of Eta_Earth? It was one of the saddest days of my life; I could only imagine how the Kepler team felt. Borucki was relentlessly upbeat, saying that the Kepler team would continue to analyze the data taken so far and would undoubtedly find more and more exoplanets, but it was clear that Kepler would no longer be able to take the exquisitely precise data that would enable it to find more Earth-like planets than could be found in the existing data.

The loss of the third reaction wheel and the definitive end of Kepler's high-precision data collection changed everything for the mission's future plans. On May 17, 2013, Grunsfeld decided that the Second Kepler Science Conference should be held at Ames in November after all, to celebrate Kepler's achievements. The conference would effectively be Kepler's funeral wake. The process of seeking a waiver from NASA HQ to allow the wake to be held began; Grunsfeld could not issue the required waiver himself. This was a 9th-floor decision at HQ. Charlie—what do you say?

You May Proceed: The Science Definition Team charged with considering using an NRO telescope for WFIRST released their study on May 23, 2013, and their Final Report was accepted by NASA. Charlie Bolden directed agency officials to continue to figure out how these unexpected assets could be used to accomplish the goals of WFIRST. Compared to the previous WFIRST design studies, which considered telescope diameters no larger than 1.3 m, a 2.4-m telescope would be considerably more powerful, but only marginally more expensive, given the gift status from NRO.

Paul Hertz, the new Director of the Astrophysics Division, noted that it might be possible to add a coronagraph to a WFIRST 2.4 m, creating a direct-imaging capability for nearby exoplanets. However, this possible goal was listed as fifth in line in the Final Report, meaning that when money ran short, as so often happens in new space missions, that goal was likely to be "descoped": the polite term for terminated. On the other hand, the sixth goal was listed as possible overlap with JWST, if WFIRST could be launched soon enough after JWST went to work. Given the prime importance of JWST, this could only strengthen the case for building WFIRST as soon as possible.

There was bad news too: the plans for ExO had been dropped. NASA announced on May 4, 2013, that only one of the two NRO telescope systems would be considered for future use, the one that would become WFIRST. Grunsfeld stated that none of the concepts emerging from the Huntsville competition were high priorities in the latest Decadal Surveys, and hence NASA would not pursue those options. Even the Astro 2010 top-ranked, WFIRST NRO option would have to wait for JWST to finish up before serious work could begin.

WFIRST was estimated to have a cost in the range of $1.5 billion to $1.7 billion, bracketing the $1.6 billion cost assumed by Astro 2010.

Soon thereafter, NASA announced on May 27, 2013, that it was proposing to divert more funds from the Planetary Sciences Division in order to support JWST and the Earth Sciences Division. The firewalls that had protected NASA Divisions from each other would be thoroughly destroyed by this operating plan. The PSD would "contribute" $44 million to JWST in FY 2013, thank you very much, if Congress did not object. However, California Representative Adam Schiff, whose turf included JPL and who was a member of the House CJS Appropriations Subcommittee, objected that this plan would "disregard the expressed will of Congress with regards to planetary science." Schiff would push back.

8

Bring Out Your Dead

I'm not dead!... I think I'll go for a walk.
—"Monty Python and the Holy Grail," 1975

The Kepler spacecraft was crippled but still very much alive and kicking. The Kepler team announced 503 new exoplanet candidates on June 14, 2013, bringing their total to 3,216, proving that the Kepler team could continue to mine the existing data set for new planets even if no more high-precision data was taken. While only 132 of these candidates had been confirmed by one means or another, the estimate was that about 90% of the candidates should turn out to be true exoplanets.

The Kepler engineers at Ball Aerospace were still working on the failed reaction wheels, trying to find a command to transmit to the spacecraft that might salvage the situation. They would continue their testing throughout the summer, encouraged by their ability to get both failed wheels to spin, at least a little, though frictional spinning would produce vibrations that could not be tolerated. Take your time, no rush; the science team had 4 years of data to process and sift through for unseen worlds, and they were confident that they could still get a good estimate of Eta_Earth with the existing data. The engineers finally gave up on August 15, 2013: the frictional vibrations were too large. Paul Hertz announced that it was time to think about what Kepler could do with just two reaction wheels. All good ideas were welcome.

On June 24, 2013, it was announced by the French and ESA that the broken CoRoT satellite was retired and would be decommissioned. CoRoT's control jets would be fired one last time, lowering the satellite's orbit into the Earth's atmosphere, where it would be left to spiral downward and burn up through frictional heating. CoRoT will perish in a circle of flame worthy of the closing act of a Wagnerian opera.

Kepler would have an altogether different fate, as it was doomed to circle the Sun endlessly, waiting patiently for about 5 billion years, waiting for the Sun to expand as a red giant and consume the spacecraft.

Game, set, match, Kepler.

This Is Getting Boring . . . Not!: Three more HZ super-Earths were announced on June 25, 2013, by a team led by Guillem Anglada-Escude (see Figure 8.1) and Mikko Tuomi. Guillem was now at the University of Göttingen, having previously spent three delightfully productive years as a Carnegie Fellow at DTM with Paul and me. Guillem's team had found the first instance of three potentially habitable super-Earths around a single star, known as Gliese 667C, itself a member of a triple-star system. Gliese 667C was on a wide orbit about the binary composed of Gliese 667A and 667B, leaving plenty of room for planets to form around Gliese 667C, a red dwarf star with a third of the Sun's mass. In fact, the team found indications of at least five, and possibly seven, exoplanets hiding in the Doppler data, which had been taken by HARPS-S, HIRES, the Ultraviolet and Visible Echelle Spectrograph (UVES) on one of ESO's VLT telescopes in Chile, and the Planet Finding Spectrograph (PFS) on Carnegie's Magellan telescope at Las Campanas.

The direct-imaging folks were scoring points as well. On June 3, 2013, ESO had announced the discovery of a gas giant planet orbiting at a considerable distance from its parent star, called HD 95086. The star was thought to be no more than 17 million years old, which meant that an exoplanet of that age and the measured brightness should have a mass about four or five times that of Jupiter. The discovery had been made in near-infrared light with one of ESO's VLTs. The gas giant orbited at a large distance from its star, about 56 times the Earth–Sun distance or twice Neptune's distance from the Sun.

Japanese astronomers made a similar announcement on August 6, 2013: near-infrared imaging had detected a gas giant four times as massive as Jupiter orbiting

Figure 8.1 Guillem Anglada-Escude, the leader of the effort that proved the existence of Prox Cen b, the closest Earth-like planet to the Solar System (Courtesy of DTM, Carnegie Institution for Science).

the young star GJ 504. The exoplanet was found at a distance from GJ 504 of 43 times that of the Earth from the Sun. The Japanese had used their 8.2-m Subaru telescope (named after the Japanese word for the Pleiades star cluster, not the Japanese auto company) on Mauna Kea in Hawaii, along with a sophisticated adaptive optics imaging system, in order to find the exoplanet. Oddly enough, GJ 504b is a pink-colored giant planet.

These two new gas giants on wide orbits, like those seen earlier around HR 8799, suggested once again that core accretion might not be the only means for forming gas giant planets; disk instability seemed more likely to be the case for forming such distant objects in situ.

It was obvious that ground-based observatories were continuing in their relentless quest to beat NASA in the race to image Earth-like planets. While these large ground-based telescopes could image pink gas giants, the hope was that the next generation of even larger telescopes (GMT, TMT, and E-ELT) would be able to photograph the pale blue dots that demarcate Earth-like worlds. Alpha Centauri B, a member of the closest star system, had already been found by Doppler spectroscopy to have an Earth-mass planet, and European astronomers had announced that they believed that this closest Earth-mass planet could be photographed by the E-ELT, an astounding claim indeed.

Exoplanet Imaging Is Back in the NASA Ball Game: On July 18, 2013, I got a phone call from the JPL scientist in charge of technology development for direct-imaging space telescopes. He inquired if I would be willing to chair a TAC that was being assembled in order to provide an analysis to NASA HQ about which particular types of internal coronagraphs should be considered for WFIRST, which was now being referred to as WFIRST-AFTA, with the AFTA standing for "Astrophysics Focused Telescope Assets," that is, the NRO 2.4-m telescopes.

I accepted the request without hesitation. This was getting more and more interesting. I had been serving for several years on a similar TAC that reviewed progress in developing the technology for ill-defined future space exoplanet imaging telescopes, but the difference here was that WFIRST-AFTA was next in line after JWST, and it had the blessing of Astro 2010. It might therefore actually fly one day. Way cool.

The WFIRST-AFTA coronagraph committee would have to make a decision by December 2013, meaning that NASA was in a rush to figure out the best way to get the NRO 2.4-m telescopes to do not only the Astro 2010 WFIRST science of dark energy and microlensing but to start looking for nearby exoplanets as well. The TAC would be charged with monitoring the progress of the chosen coronagraphic technology once the new program started in earnest in 2014.

On top of this positive breakthrough, NASA HQ finally decided to allow the Second Kepler Science Conference to be held as planned, in spite of the ongoing federal budget crisis and the cancellation of other NASA meetings. Kepler would have its swan song after all. Kepler had proven that Earths were commonplace, and it would go out with a celebratory conference at NASA Ames, where it was born and nurtured.

Borucki Wins an "Oscar": In a White House ceremony on October 24, 2013, President Barack Obama awarded Bill Borucki with a "Sammie," the equivalent of an "Oscar" award for federal employees, in honor of the incredible success of his Kepler planet-finding mission. More honors were certain to be headed Bill's way that would be considerably more well-known (and lucrative) than the Samuel J. Heyman Service of America Medal. Still, it is not every day that an astronomer gets to shake the President's hand and receive a medal for work as fantastic as Bill had accomplished with Kepler. A second Sammie was awarded to the NASA team responsible for the Mars Curiosity rover.

A pattern was emerging here: the search for life beyond Earth was considered to be great stuff by the White House, it seemed. It would be even better if NASA's science budget reflected that enthusiasm; the FY 2014 President's budget request had flat-funded NASA's SMD for the foreseeable future. Given the federal budget deficit, though, perhaps that was the best that even a supportive Administration could afford.

The Second Kepler Science Conference began on schedule on Monday, November 4, 2013, with the press conference announcement by Natalie Batalha that Kepler had found 3,553 planet candidates in just the first 34 months of data, with another year of data remaining to be analyzed. Over 600 of the candidates had radii no larger than that of Earth, and over a hundred orbited within their star's HZ. Two dozens of the HZ planets had Earth-like or smaller sizes. Half of the red dwarf stars had HZ planets with sizes less than twice that of Earth. Red dwarfs were not the primary targets for Kepler, with only a few thousand of them being included in the 150,000 or so stars being monitored for transiting planets, but they were turning out to be fertile targets for exoplanet hunters.

Bill chaired the Monday afternoon session on exoplanet statistics, where various estimates were presented for HZ super-Earths, ranging from 22% for sun-like stars to 50% for red dwarfs. The next day, Wes Traub announced his refined estimate for Eta_Earth was now 88%. Curiously, there were talks by Kepler team members but still no official Kepler team estimate of Eta_Earth. Still, the estimates from team outsiders were convincing enough to make the case for what Kepler would find when all the data was processed and adjusted for observational biases.

Charlie Bolden phoned in with his congratulations for Bill Borucki and the Kepler team on Tuesday afternoon. Clearly Borucki deserved his Sammie, even if the Kepler team was not yet ready to announce their best guess for η_E, the factor in the Drake Equation that was the primary goal of Kepler. In fact, Frank Drake himself gave a public lecture to an overcrowded room that evening, where he offered his answer to Fermi's paradox. Enrico Fermi, the famed nuclear physicist, had wondered in 1950, *"Where are they?"* in response to dreamy visions of a universe rife with intelligent life. Kepler's results were pointing toward a universe riddled with potentially habitable worlds (i.e., η_E closer to unity than to zero), so Fermi's question had acquired new-found urgency. Drake pointed out that the amount of energy required to accelerate a spacecraft to anywhere close to the speed of light is ridiculously huge (a direct consequence of Einstein's theory of special relativity), so there was no need to worry about Klingons or other unwanted interstellar visitors arriving in Earth orbit. Stephen Hawking could sleep better now.

Europeans Throw Long: ESA launched the Gaia space telescope from Kourou, French Guiana, on a Soyuz-Fregat VS06 rocket on December 19, 2013, beginning a 5-year mission to map the positions of a billion stars on the sky to unprecedented precision, at the bargain-basement price of about one euro per star. Gaia would head out to an L2 orbit, on a line with the Earth and the Sun, where it would be joined by JWST in about 5 more years.

Gaia would concentrate on searching stars within about 650 light-years (200 parsecs) of the Solar System for any unexplained apparent motions beyond those caused by Gaia's orbit around the Sun and the motion of the target stars across the sky. The focus would be on relatively bright stars, similar to the Sun; most red dwarf stars would be too faint. Any stars that seemed to wobble around would be accused of harboring Jupiter-mass planets responsible for the stellar wobbles. Gaia expected to find evidence for about 1,000 such exoplanets, though they would have to be on considerably shorter-period, closer-in orbits than our Jupiter's 12-year orbit in order to see a complete wobble in 5 years. Following a star's wobble over a complete orbital period is the most basic criterion for the reliability of an astrometric planet detection.

Despite the head start given by Peter van de Kamp's visionary exoplanet search, begun in 1938, ground-based astrometry had yet to discover a single reproducible exoplanet. Gaia was going to try to be the first, though only by going to space to avoid that pesky Earth atmosphere's effect on the "seeing." As a result, with 70 epochs of observations, Gaia was hoping to measure stellar wobbles with an accuracy of 24 microarcseconds, about 100 times less accurate than SIM would have accomplished, but still a good factor of 10 better than had been done from the ground. Gaia would certainly find something interesting, even if it could not find exoEarths.

Gaia would be ESA's most expensive space telescope capable of discovering new worlds for many years to come, as the head of the agency had announced on October 31, 2013, that the next two large-class space missions, L2 and L3 (L1, or JUICE, would head to Jupiter's icy moons), would be devoted to X-ray observations (Athena+) and to gravitational wave detections (Evolved LISA, eLISA), with launch dates of 2028 and 2034, respectively. The European exoplanet community noted that the competition for the next chance for a major ESA exoplanet mission, L4, would begin in another 6 years, with a target launch date of 2040, and soberly noted that the younger members of the community would have to carry the responsibility of accomplishing such a distant mission.

ESA had effectively dropped out of the space race to directly image exoplanets. The European exoplanet fanatics were left with hoping that NASA might invite their participation in such a flagship mission, much as had happened with HST and JWST. Sitting in the back seat on a flagship mission was better than being left behind altogether.

Bolden Throws Up His Hands and Surrenders: Charlie Bolden told the NAC Science Committee on December 4, 2013, that the days of flagship missions at NASA were over: NASA's budget could no longer support such billion-dollar-plus missions. Charlie was tiring of the annual battles with the OMB to maintain support for flagship missions, and he suggested that scientists consider proposing smaller, more frequent missions. JWST might then turn out to be one of the last of its kind. If so, there might not even be a back seat for the Europeans to climb into.

Up on Capitol Hill, a few blocks to the east, a House Committee was talking on the same day about future flagship missions to possible abodes of life in our Solar System, such as Jupiter's Europa and Saturn's Enceladus, both icy worlds with possible subsurface liquid water oceans and aquifers. The Committee chair, Lamar Smith, joined with John Culberson and Adam Schiff to voice their support for a Europa Clipper mission, estimated by JPL to cost about $2.1 billion.

On December 13, 2013, I joined a group at NASA HQ that was presenting the results of the committee I had chaired that was charged with overseeing the decision about which coronagraphic designs should be aggressively developed in hopes of removing any remaining technology hurdles in time for a coronagraph to be included in the WFIRST-AFTA space telescope. Several months of intensive studies had shown that even with the non-optimal, obscured pupil of the NRO telescope assemblies, properly designed coronagraphs could largely overcome that hurdle and be able to image a fair number of the then-known exoplanets already found by Doppler spectroscopy. Six different coronagraph designs had been studied to death, and the end result was that several concepts looked promising from the point of view of factors such as technological readiness, science merit, and development cost.

With a total cost estimate of about $2.1 billion, WFIRST-AFTA was considered a flagship but with a cost less than one-fourth that of JWST. Paul Hertz, the Astrophysics Division Director, gave us the all clear to proceed with planning for the AFTA coronagraph.

I would soon be asked to chair not only the coronagraph technical oversight committee but also the WFIRST infrared detectors committee, reviewing development of the Hawaii 4RG detectors that would enable the wide-field infrared imaging that was the primary goal of WFIRST, before WFIRST became WFIRST-AFTA. How could I say no? This was likely to be the only direct-imaging exoplanet ride for a long time to come, and I had been left waiting at the station for one since 1988. I wanted a seat too, before I became too senile to appreciate the results. WFIRST-AFTA it is, though Europeans were not being offered tickets.

Kepler Catches a Break: On Saturday, January 4, 2014, I was feeling ill, so I sat by myself in the back of the room at the ninth meeting of the ExoPAG, being held in the Gaylord National Resort Hotel in National Harbor, Maryland. I learned that Kepler had been extremely fortunate in the way that it had been defeated by mechanical problems.

The failure of two of Kepler's four reaction wheels, used for pointing the telescope, had ended Kepler's prime mission prematurely. However, we learned that Kepler was not dead yet. By sheer luck, the two failed reaction wheels both controlled pointing in the same direction, along the telescope optical axis, leaving the remaining two able to point Kepler accurately in a given direction, but letting it spin about the line of sight in that direction. That was bad. But with a clever plan for pointing Kepler in such a way as to minimize the torque on the telescope derived from the Sun's radiation, which would spin up the telescope around its optical axis, Kepler could point accurately for a few months at a time. The key was to limit Kepler to searching fields of view that lay in the same plane as that of Kepler's Earth-trailing orbit around the Sun: when oriented at just the right roll angle, the net torque from the Sun's radiation would be zeroed out.

With this trick, Kepler would be able to continue hunting for exoplanets with its photometric precision degraded by a significant but acceptable amount. The hope was to achieve a precision of about 90 parts per million, about 4.5 times worse than with all four reaction wheels working. Not bad, not bad at all, considering this was still much better than could be done from the ground.

The Kepler team proposed the "K2" mission to NASA HQ as a follow-on to the Kepler Mission, a double pun that captured the secondary nature of the K2 mountain peak only to Mount Everest itself. The proposal listed 10 fields in the ecliptic plane where K2 would stare for 80 days at a time, sufficient to find HZ super-Earths around red dwarfs, stars that had surpassed sun-like stars in importance in the search for life beyond Earth. A first test of this new approach to keeping the spacecraft able to detect exoplanets was run successfully in October 2013. The K2 campaign would begin in May 2014, and was slated to run to May 2016.

In the meantime, the Kepler team announced a near-doubling of the number of confirmed exoplanets on February 26, 2014. A total of 715 exoplanet candidates had been confirmed as true exoplanets by virtue of the fact that all 715 candidates were in multiplanet systems, where more than one transiting planet had been found. While background eclipsing binary systems could masquerade as transiting planets, the chances of having more than one such charlatan creating a false signal for a multiplanet transiting system were small enough to be negligible. Kepler was still turning out exoplanets, with more to come.

Transits, Transits, and, Surprise, More Transits: Given the modest success of ESA's CoRoT Mission, and the overwhelming success of Kepler, ESA decided on February 19, 2014, to proceed with the proven technique of transit planet detection and planned to launch yet another transit search mission. The PLATO mission would be ESA's third medium-class mission (M3), scheduled for launch in 2024. PLATO would have 34 small telescopes pointing in different directions, allowing it to scan roughly half of the sky, staring at the closest, brightest stars, looking for HZ rocky planets with longer-period orbits than NASA's TESS would start seeking in 2017. PLATO would scrutinize about a million stars from its orbital station at L2, examining two fields, well above and below the ecliptic plane, the opposite of where K2 would be looking.

In October 2012, ESA had announced plans to launch CHEOPS in 2017, its first small mission (S1), in partnership with the Swiss exoplanets group. CHEOPS would be a clone of CoRoT and hence inexpensive to build, with a 30-cm mirror, compared to 27 cm for CoRoT. CHEOPS would perform follow-up transit photometry of exoplanets discovered by ground-based searches around nearby bright stars.

It seemed that transits had become the norm for exoplanet space telescopes: CoRoT and Kepler would be followed by CHEOPS, TESS, and now PLATO. But none of them would take an exoplanet's picture, and so far no exoplanet had bothered to take a selfie and send it to Earth, at least not so far as the SETI folks could tell.

Ground vs. Ground vs. Ground: The battle of the next generation of giant ground-based telescopes continued unabated, with much of the struggle focused not on the sky but on raising partners and funds. By March 2014, it was estimated that the GMT project had raised over half of the needed $800 million, the TMT folks had raised about 80% of their $1.2 billion, and the E-ELT had about two-thirds of their $1.5 billion committed. All three projects nominally were ready to start "construction" in the spring of 2014, with first light expected in 2020, 2022, and 2024 for the GMT, TMT, and E-ELT, respectively.

Oddly enough, Brazilians were split between joining GMT and the E-ELT: Brazil was planning to join ESO as a nation, and therefore the E-ELT, but the Brazilian state of São Paulo was planning on joining GMT, apparently in part as a result of the severe difficulties that the Brazilian national government was facing in securing the funds necessary to join ESO. Brazil was expected to cough up nearly $400 million to join ESO, in spite of not having nearly as many astronomers as the other ESO nations, and even some Brazilian astronomers objected to such an expensive membership. Major construction contracts for the E-ELT could not be awarded until Brazil made up its mind. The state São Paulo Research Foundation decided to move ahead on their own and consider a proposal to invest $40 million in the GMT, a 4% share that would yield 4% of the GMT observing time, to be paid for by the São Paulo taxpayers.

By chance, I had noticed a headline in the *West Hawaii Today* newspaper on April 13, 2013, when in the Kona airport after a scientific meeting on the Kailua-Kona coast. The headline screamed "30-Meter Telescope approved": the Hawaiian Board of Land and Natural Resources (BLNR) had granted approval for constructing the TMT on Mauna Kea. The article noted that both native Hawaiians and environmentalists had petitioned against the project, but the Board approved the project nevertheless, provided that the TMT employees were trained in cultural and environmental awareness. No problemo. We've got this.

The site for GMT, Cerro Las Campanas at the Las Campanas Observatory in Chile, had already been leveled by the extensive use of explosives on the hard granitic rocks that formed the rugged mountaintop. While observing at Las Campanas in 2012, we would gather after lunch to watch the day's explosives go off, a few miles away from the LCO lodge and dining hall. A satisfying plume of dust would arise, and then we would head off to the domes to get ready for the evening's observations.

Existing ground-based telescopes in Chile were continuing the battle to detect exoplanets by direct imaging. The newly built Gemini Planet Imager (GPI) on the Gemini South telescope on Cerro Tololo was operational and an image had been published in *Science* of the gas giant planet orbiting the famous debris disk star Beta Pictoris on February 21, 2014. The SPHERE (Spectro-Polarimetric High-contrast Exoplanet REsearch) coronographic imager was slated to begin operations in May 2014 on an ESO VLT telescope on Cerro Paranal. These instruments would use adaptive optics to minimize the effects of Earth's atmosphere and were expected to rule the roost for such studies until the advent of the giant telescopes in the 2020s.

WFIRST-AFTA Dodges a Bullet: On March 18, 2014, the NAS released its findings about the idea of adding a coronagraph to WFIRST-AFTA, noting that although the coronagraph would do wonders for exoplanet imaging, the technology needed to fly the coronagraph was clearly not flight ready. The NAS report recommended that NASA move aggressively to advance the technological readiness of the coronagraph option to acceptable levels, allowing its inclusion when WFIRST-AFTA was ready to be formally proposed as a new start for the next NASA Astrophysics Division flagship after JWST. The coronagraph technology development committee that I was chairing for WFIRST-AFTA was busy assessing the rapid progress already being made. We all intended to be ready when the time came for a new start.

In spite of the coronagraph not having been a feature of Astro 2010's WFIRST concept, the NAS committee had evidently seen the opportunity to move the field of exoplanet imaging along much faster than would otherwise be the case. They had done the right thing, in my biased opinion, something that is not always the outcome for committees composed of competing factions.

The NAS report also noted that using the donated NRO optics would raise the risk of increased cost growth for WFIRST-AFTA, as it was unclear what modifications might be necessary for it to point up instead of down. The President's Budget Request for FY 2015, revealed on March 4, 2014, had kept NASA's top line flat; but $14 million was earmarked for WFIRST-AFTA. Score! JWST's request was $645 million, as agreed in the 2012 replan. Steady as she goes, skipper, but look out for the wake.

Close, But No Trip to Stockholm Yet: The Kepler team continued to roll out tantalizing discoveries of nearly Earth-like planets. A *Science* paper on April 18, 2014, revealed the parameters of Kepler-186f, an exoplanet with a radius just 10% larger than that of Earth, orbiting in the HZ of its star.

Unfortunately, the Kepler-186f host star was a red dwarf, considerably lower in mass than the Sun, so Kepler-186f could not be considered as a true Earth 2.0. Other Kepler discoveries had also come close but no cigar: Kepler-20e was 20% smaller in size than Earth, and orbited a sun-like star, but it was too hot to be habitable; and Kepler-22b orbited in the HZ of a sun-like star, but it was over twice the size of Earth. Kepler was continuing to find that multiple planet systems were common: Kepler-186f was the fifth planet discovered in that system alone, and others might be lurking there as well.

A Bargain Indeed: The WFIRST-AFTA project released an interim report on May 6, 2014, the work of Spergel's and Gehrel's Science Definition Team, which stated that the cost of adding a coronagraph to the baseline mission was estimated to be $260 million. Compared to a baseline cost of about $2.1 billion, the added capabilities that a coronagraph would bring to WFIRST-AFTA made an added $260 million seem like a deal that could not be ignored. Adding the ability to image exoplanets could only enhance the attractiveness of WFIRST-AFTA for those humble souls who cared more about life beyond Earth than about baryonic acoustic oscillations. Congressional people in particular would find that adding the coronagraph to WFIRST-AFTA strengthened their support for what might otherwise seem to be a rather esoteric mission.

Several relatively low-cost, "probe-class" missions, with total costs less than $1 billion, were also being studied by NASA: the "Exo-C" coronagraph and the "Exo-S" star shade, and their necessary technology developments, had been assessed. Exo-C's cost estimate did come in at less than $1 billion, but Exo-S was estimated at $1.6 billion, though you could select just the star shade for a mere $600 million in this limited-time offer . . . just pay the separate shipping and handling costs. Getting a coronagraph on the 2.4-m WFIRST-AFTA for $260 million seemed like quite a bargain in comparison. Exo-C and Exo-S could be held safely in reserve, pending the fate of WFIRST-AFTA's new start in 2017 or so.

The K2 mission proposal won approval by NASA HQ's Senior Review on May 16, 2014, where all space missions compete for extensions once their primary goals have been achieved. The Kepler mission extension proposal was also accepted by the Senior Review, though a proposal to extend Spitzer was not accepted. K2 would keep part of the original Kepler team going for another 2 years, that is, until the next Senior Review rolled around, while the rest of the team finished the analysis of the Kepler data in hand.

Keeping Kepler and K2—that is, TPF-T and TPF-T2—running would cost NASA about $20 million a year, another bargain that could not be ignored.

SMD decided to keep Spitzer running for 2 more years, overruling the Senior Review recommendation. The Spitzer Warm Mission was doing great stuff with transiting exoplanets, even though its supply of liquid helium cryogen had run out in 2009, rendering it useless for long-wavelength observations. Two short-wavelength channels were still functioning, at 3.6 microns and 4.5 microns, perfectly fine wavelengths for studying the atmospheres of transiting hot Jupiters and hot super-Earths.

What a Blast: On June 19, 2014, ESO tried to catch up with GMT in the mountain-shaping business by beginning to blast away the top of Cerro Armazones in order to make room for the construction of the E-ELT. Cerro Armazones is located in Chile hundreds of miles north of the LCO, but only about 15 miles from the four existing VLT telescopes on Cerro Paranal, close to the northern Chilean city of Antofagasta.

Although blasting had begun, Brazil's national government had still not ratified their 2010 agreement to join ESO. Brazil was expected to contribute about a quarter of the E-ELT cost, but until Brazil signed in ink, the E-ELT project would be in stasis. In a country like Chile, where hard-rock mining is a principal industry, procuring explosives and the services of those who know how to use them is considerably less expensive than building a dome capable of housing a 39-m diameter telescope, much less the telescope itself.

A top ESO official remained confident that Brazil's national government would sign the deal for ESO membership, though beads of sweat were probably beginning to appear on his brow. The Brazilian membership fee appeared to have been cut in half, compared to what it was reported to cost a month earlier, which had to help grease the skids for the deal in the Brazilian National Congress.

The São Paulo Brazilians, however, moved ahead in July 2014 and joined GMT, with a $40 million contribution, joining the seven U.S. partners as well as two Australian partners and South Korea. All right, all right, all right.

Canada, China, India, and Japan had joined the TMT project, with the plan being that their contributions would be to develop the technology needed to build various TMT components rather than to simply provide funds for the project. Canada alone had committed to spending over $243 million on the TMT, a sizeable fraction of the total cost. China in particular viewed the TMT as a means to improve its technological prowess—forward engineering is better than reverse engineering. In fact, China was hedging its bet on the TMT by planning on developing and building its own large telescope, a 12-m one that would give Chinese astronomers bragging rights for having the largest telescope on the planet, at least until the next generation of extremely large telescopes began their work. But first they would have to figure out how to build a state-of-the-art, 12-m telescope for an estimated cost of about $220 million, somewhere in the mountain ranges of western China. Presumably the citizens of that remote region, close to the borders of several former Soviet Union republics, would not object to the ambitious project.

Wait Just a Minute: A ground breaking ceremony for the start of construction for the TMT on the Big Island of Hawaii's Mauna Kea was planned for October 7, 2014. This time, no explosions were planned to level the site, as the top of Mauna Kea is largely composed of unconsolidated volcanic materials, more akin to a gigantic cinder pile than the granitic mountains of the Atacama in northern Chile.

A group of native Hawaiians and non-Hawaiians also made the long journey to the top of Mauna Kea, not to join in the TMT celebration but to protest the groundbreaking for yet another observatory on top of the mountain that is the most sacred to native Hawaiians. This could not have come as a surprise: TMT had already been delayed for several years by protests about adding TMT to the Mauna Kea Observatory (MKO). Nevertheless, the groundbreaking ceremony was completed, and plans continued for starting ground preparation for the telescope. The TMT cost estimate had risen to $1.4 billion, 80% of which was committed by the TMT partners, largely as in-kind contributions rather than as cash transfers, which was the GMT business model.

WFIRST-AFTA Gets Serious: The first general science meeting in support of the newly envisioned WFIRST-AFTA space telescope was held in the Pasadena Hilton Hotel, starting on November 17, 2014. On that morning, David Spergel led off with a summary of the mission, noting that it was now in "pre-Phase A," with $66 million committed to its development. WFIRST-AFTA was being made ready for a fast ramp-up once it entered Phase A, possibly leading to a launch as early as 2023. I followed with a talk about the status of exoplanet research in the wake of the surprises discovered by Kepler and in anticipation of the surprises to be expected from WFIRST-AFTA, with a focus on how WFIRST-AFTA would permit the direct imaging of considerably fainter exoplanets, compared to their stars, than would be possible with even a gigantic ground-based telescope.

The next day, Paul Hertz gave us the view from NASA HQ, which was purely good news. JWST was still on schedule for a launch in October 2018, and on budget as well. Best of all, the plan was for the APD to retain the JWST funding wedge once JWST launched, meaning that we could afford to build and launch WFIRST-AFTA. Equally important, Congress was overruling the President's NASA budget priorities and had inserted significant funds dedicated to WFIRST-AFTA development for the 2nd year in a row. That meant that coronagraph development work could continue unabated. Hertz was hoping for a new start for WFIRST-AFTA in FY 2017, which would begin in less than 2 years, on October 1, 2016. All systems were go for launch in 2023.

Poland Gives the E-ELT a Boost: With the decision of Poland to join the project, coupled with what can only be called some creative accounting, the head of ESO announced on December 4, 2014, that enough funds had been secured to proceed with the construction of the first phase of the E-ELT, a somewhat stripped-down version of the original concept, with the fate of the second phase resting on the addition of further partners. The second phase would include the remaining one-quarter of the 798 mirror segments making up the primary mirror, and part of the adaptive optics system, which would be needed to make the sharpest possible images of nearby exoplanets. This meant that the E-ELT would begin life blind to exoplanets unless funding for the second phase could be secured. The press release hinted obliquely at this by stating that the E-ELT would "allow the initial characterization of Earth-mass exoplanets." Exoplanet masses meant that the E-ELT would start out with a whiz-bang Doppler spectrometer but no direct-imaging capability for these sought-after worlds.

The E-ELT was shooting for first light in 2024, just a decade away in 2014. The total cost of the E-ELT was estimated at $1.34 billion, down somewhat compared to the previous $1.5 billion estimate, largely as a result of changes in the exchange rate between euros and dollars. No mention was made of the Brazilians in the ESO announcement. In this arena, no news is generally bad news. What would be the source of the rest of the money needed to allow the E-ELT to image nearby Earths?

GAO Weighs In: On December 16, 2014, the GAO released a report that was critical of JWST, stating that although the JWST project reported that it was still on schedule and on budget, the remaining schedule reserve was dropping as a result of delays on every single major subsystem during the past year, putting the entire project at some risk. The GAO stated that the "cost risk analyses used to validate the JWST budget are outdated" and recommended that NASA HQ should follow "best practices when updating its cost risk analysis to ensure reliability": ouch. If JWST did not launch on time, it was not likely to launch on budget, and that would be bad news for WFIRST-AFTA. Those of us hoping to launch WFIRST-AFTA had likely become JWST's strongest supporters.

The Omnibus Budget agreement for FY 2015 had been released the previous day, including the promised $50 million reserved for WFIRST-AFTA. Thanks to the support of Representative Frank Wolf and Senator Barbara Mikulski, NASA SMD got a $93 million raise instead of the $179 million cut proposed by the White House. A total of $100 million was earmarked for the Europa Clipper mission, and Mars research got a boost as well, so the search for life beyond Earth was still a compelling vision for Congress. Wolf was stepping down from Congress and would be replaced as chair of the House CJS Appropriations Committee by John Culberson, a strong supporter of the Europa mission, who stated "if anybody tells me about a hole in NASA's budget, I will plug it." Thanks—we'll be in touch.

9

Ominous Signs from Maryland

Time is what prevents everything from happening at once.
—John Archibald Wheeler, 1978,
American Journal of Physics, 46, 323

On March 2, 2015, Maryland Senator Barbara Mikulski announced that she would not seek re-election to a sixth term. Mikulski was the longest-serving woman in Congress, having first been elected to the House in 1976. Senator Mikulski's role as the chair or ranking member of the Senate Appropriations Committee and the CJS Appropriations Subcommittee meant that she had had a disproportionate influence on protecting NASA's budget, or at least those elements based in her native city of Baltimore (i.e., STScI), or nearby (i.e., GSFC). NASA would have 2 more years of her strong support; but after that, NASA could no longer expect to have such an influential voice on its side in the Senate budget battles.

On March 17, 2015, I was back at NASA HQ, serving a second round on the Astrophysics Subcommittee of the NAC, despite having been kicked out as chair of the APS several years earlier as a result of my seemingly intemperate remarks comparing JWST to a certain Gulf of Mexico hurricane. I was pulled aside at this first meeting and warned to keep my mouth shut this time—if I did not, those who had argued to put me back on the APS would be in trouble. John Grunsfeld would be watching me closely.

The APS meeting had a presentation about the status of JWST, where the mantra was repeated—"on schedule, on budget"—and it was noted that by FY 2019, the JWST budget wedge could be largely devoted to WFIRST-AFTA. When it came time for comments from the APS, I managed to suppress my strong desire to mention casually the risk associated with the plan to launch JWST from French Guiana in October 2018, right in the middle of the Northern Hemisphere hurricane season. I chuckled to myself, but let it go this time. It might come in handy later.

I had been added to the APS because NASA wanted me to chair yet a fourth committee at the same time as the three others I was chairing, namely, the APS's ExoPAG, whose chair was a member of the APS by definition. The ExoPAG was about to be charged by the APD with helping to decide which flagships should be planned to follow JWST and WFIRST-AFTA, perhaps to launch sometime around 2030. The Astro 2020 Decadal Survey could be expected to set priorities for the 2020–2030 time period, and that survey would be getting started around 2019.

The NASA APD needed to begin serious consideration of possible flagship missions for consideration by Astro 2020, and now was the time to spend a few millions of dollars performing detailed studies of what should be ranked by the Survey. One likely candidate mission was being called HabEx—the Habitable-Exoplanet Imaging Mission. HabEx would avoid the obscured pupil forced on WFIRST-AFTA by the generous donation of the NRO optical assemblies; and as a result, HabEx should be able to detect and study considerably more exoplanets than even WFIRST-AFTA could, for example, real Earths rather than mere super-Earths. It was looking like time to dust off the old plans for TPF-C and to think about how to improve that now-ancient design in the context of the HabEx possibility.

Grunsfeld was pushing for a space telescope even grander than TPF-C: a concept called ATLAST, the Advanced Technology Large-Aperture Space Telescope, with a diameter in the range of 8 to 16.8 m. The ATLAST concept had been developed by STScI as a successor to Hubble—it would certainly guarantee full employment at the Institute and at parts of GSFC for decades to come if it was accepted and built as planned by 2025 to 2035. During a hallway conversation at the APS meeting, I had mentioned casually to someone who I did not know at the time was working on ATLAST that I did not think that NASA could afford a 16.8-m telescope, and this new intemperate remark was quickly reported to Grunsfeld. Yikes. Watch what you say when speaking your mind at a NASA HQ advisory committee meeting, even in the hallway.

Did You Not Hear Us the First Time?: Protests began again in earnest at the summit of Mauna Kea on April 2, 2015, with about 100 people blocking the summit road and trapping TMT construction workers for over 8 hours. Police arrested several dozen protesters, but the protests continued and widened. On April 13, hundreds of University of Hawaii at Manoa faculty members and students walked out of their classes, and thousands more on Oahu joined in protest. The International Indian Treaty Council weighed in as well, calling for international action against the TMT construction plans and the establishment of a "Hawaiian government with all the powers of an independent state" in order to halt the "cultural genocide of Mauna Kea."

On May 26, 2015, the Hawaiian governor announced that TMT could proceed as planned but that 25% of the dozen or so MKO telescopes would have to be removed in the next decade in deference to the protestors. A promise was made that the TMT would be the last telescope built on Mauna Kea. The University of Hawaii, which leases the MKO land from the state of Hawaii, began planning for the removal of the required 25% of the telescopes in order to accommodate the TMT.

The TMT Board of Directors decided to resume construction on top of Mauna Kea on June 24, 2015, feeling that they and the University of Hawaii had sufficiently addressed the concerns of the native Hawaiians who had protested the desecration of their sacred mountaintop. However, when the TMT construction workers tried to drive to the summit on that day, accompanied by officers, they found the road blocked by hundreds of protestors and several large boulders, leading to another dozen arrests and a retreat back down the volcano. Legal challenges to the TMT were underway in Hawaiian courts, with hearings scheduled in August 2015 for a challenge to the TMT building permit.

In this situation, the TMT project folks could not decide whether, or when, to try to resume construction. The GMT and E-ELT project folks kept quiet, happy to have the strong support of the Chilean people and government for their site locations.

Borucki Awarded the Shaw Prize: On June 1, 2015, it was announced that Bill Borucki had won the $1 million Shaw Prize, often described as the Asian Nobel Prize. The Shaw Prize was endowed in 2002 by the late Run Run Shaw, a legendary Hong Kong television and filmmaker. The Royal Swedish Academy of Sciences had missed its chance to be the first to award Borucki a major prize, a real Nobel Prize. Bill invited me and a few other Kepler team members to attend the ceremonies in Hong Kong with him, and I gladly accepted. I figured that would be as close as I was going to get to being awarded such a major scientific prize. Besides, I had already been to Stockholm. Sniff.

Bill announced that he would retire from NASA on July 3, 2015, after 53 years of service, so the award of the Shaw Prize would be a fitting retirement present. The Kepler team had by then confirmed the existence of over 1,000 exoplanets, and it was still searching the data for more. The Shaw Prize had thus rewarded Bill with a prize of about $1,000 per confirmed exoplanet. With a total cost of over $600 million, NASA and the U.S. taxpayers had spent about $600,000 to find each of these new worlds, a bargain at that price.

NASA Shows Its Hand: On June 2, 2015, NASA released its agency-wide technology development plan for the next decade. Buried in the huge document was a table showing that the plan assumed that the next flagship mission to launch after WFIRST-AFTA would be something like HabEx, scheduled for launch in 2030, as expected. Several other candidates for future consideration by Astro 2020 were slated for launch 5 or more years after the "Exoplanet Direct Imaging Mission." Things were looking good for WFIRST being an important first step toward NASA's goal of imaging nearby Earth-like planets.

What's a Billion Dollars Worth These Days?: The GMT project cost estimate rose to a cool $1 billion, but a press release on June 3, 2015, announced that the 11 GMT partners had approved spending the first $500 million for GMT construction. The press release noted that the GMT would be looking for new planets, but it did not mention direct imaging. Evidently GMT would be focusing on detecting exoplanets by Doppler spectroscopy, at least to start, much as seemed to be the case for the first phase of the E-ELT.

First light was still scheduled for 2021 for the GMT, but with only four of the eventual seven mirrors in operation. The four-mirror first phase was estimated to cost about $700 million, not too much more than the $520 million already committed. Provided that sufficient additional funds were committed by new partners, the full complement of seven GMT mirrors was planned for operation in 2024, about the same time that WFIRST-AFTA would launch.

And the Answer Is . . .: Several new estimates of η_E, the fraction of stars with Earth-like planets, began to emerge in the astronomical literature, based on the pioneering results of the Kepler Mission. Given the need to extrapolate the Kepler data to the desired planet masses and orbital periods, there was still a fair amount of uncertainty on the correct value for η_E, with estimates ranging from about 0.24 to as high as 1.2. The latter estimate was truly astounding, if correct: the typical sun-like star would have at least one habitable, Earth-like planet, and some would have more than one. Mysteriously, there was as yet no formal announcement of the best guess of the Kepler Mission for η_E, which was, after all, the main reason why NASA had funded the mission. When would the Kepler Mission give us its number?

At the ExoPAG meeting in the Chicago Hilton Hotel on June 13, 2015, Rus Belikov of NASA Ames outlined his plan to start a Study Analysis Group (SAG) that would attempt to make sense out of the various estimates for Eta_Earth. I could only wish him luck and present his proposed SAG charter at the next APS meeting for formal approval.

Courtney Dressing and David Charbonneau published an article in the July 1, 2015, issue of ApJ that updated their estimates for the frequency of HZ exoEarths around red dwarfs to being in the range of 0.12 to 0.24, depending on the assumed exoplanet diameter and extent of the HZ. Wes Traub submitted an ApJ article on July 14, 2015, with his latest estimate of Eta_Earth now equal to 1.22, plus or minus 0.07 planets per star. It was impressive that Wes was quoting an uncertainty in the third digit, when others had come up with numbers differing by factors of order 10, but there you have it: Eta_Earth was a big number, of order unity. Rus would have fun trying to get everyone to agree on a single number.

Another Billionaire Picks Up the SETI Phone: On July 20, 2015, Russian billionaire Yuri Milner tossed $100 million into the search for life beyond Earth by endowing searches with radio telescopes in the United States and Australia for transmissions from extraterrestrial life. The "Breakthrough Listen" project would pay 20% of the costs of operating the giant Byrd Green Bank Telescope in return for 20% of the observing time. Milner also bought time on the Parkes Radio Telescope in New South Wales, Australia, thereby covering both the northern and southern skies.

The pioneering SETI Institute search had fallen on hard times after billionaire Paul Allen limited the amount of funds he was willing to provide to build and expand the ATA in California to a total of over $30 million. Apparently Allen got tired of waiting for the ultimate phone call from space and decided to spend more time with his professional sports teams, the Seattle Seahawks and the Portland Trail Blazers, along with a number of more worthy philanthropic efforts sufficient to be awarded an Andrew Carnegie Medal of Philanthropy in 2015. So be it. SETI Institute leader Jill Tarter had begun full-time fundraising in order to keep the power on at the ATA in Hat Creek, California, and had raised enough to run the ATA SETI search for 12 hours each night.

So now Yuri Milner had decided to join the SETI waiting game with his Breakthrough Listen initiative. Searching for life beyond Earth would be a lot easier if we just got a phone call once in a while. What, you're too busy to call? After all we did for you? Well, nothing, actually, but we would sure like a call. Please? Even 140 characters would be great. Thanks.

Can't We Build It Any Faster?: Both houses of Congress decided they wanted to see images of extrasolar planets taken by WFIRST-AFTA, and sooner rather than later. In fact, the sentiment was that the coronagraph option for WFIRST-AFTA was not merely an option, a nice frill that could be descoped once the project ran into difficulties. The word was that both the House and Senate wanted the coronagraph on WFIRST, or else, and that they would even like to see the schedule accelerated.

Paul Hertz announced at a meeting of the Astrophysics Subcommittee on July 21, 2015, that a new start for WFIRST might come as early as FY 2017, and possibly in FY 2016, with a launch date planned for 2024 or so. The Senate in particular wanted a new start in January 2016 and was calling for $90 million for WFIRST in FY 2016. Clearly, Senator Mikulski wanted to give NASA one more magnificent gift before she left the stage. I could not argue with that gesture, not at all. Thank you, Senator, thank you very much.

Hertz also mentioned that his best estimate of the APD funds expected available for "spending" by the Astro 2020 folks from the FY 2025 to FY 2035 time period was $5 billion. This reflected the expectation that WFIRST would launch by 2024 so that the WFIRST funding wedge could be devoted to new missions thereafter. The key point was that APD did not expect to be able to start anything in FY 2020, as Astro 2010 still owned the first 5 years of that next decade. This meant that Astro 2020 should be renamed Astro 2025. Still, $5 billion was a lot, perhaps enough for a flagship and a few support ships for the NASA space flotilla.

Twenty Years Old and Still Growing: The twentieth anniversary of the announcement of the discovery of 51 Peg b was celebrated with a gala event, "Exoplanets 20/20," at the Smithsonian National Air and Space Museum (NASM) on the Mall in Washington, DC, on the evening of October 20, 2015. Hundreds of astronomers and NASA officials partied, noshed on platters of hors d'oeuvres, and watched a JPL-produced movie in the NASM IMAX theater celebrating the achievements of the last two decades, even as plans were afoot for far grander (and expensive) exoplanet searches from space. The IMAX film featured interviews with pioneers Michel Mayor, Paul Butler, and Geoff Marcy, though Marcy's footage had been cut back substantially at the last minute as a result of his resignation from UC

Berkeley for sexual harassment the week before. Marcy had also resigned as the principal investigator for the Breakthrough Listen project at Berkeley, which had been announced just 3 months earlier.

The next evening, October 21, 2015, Paul Butler, Kepler's Natalie Batalha, and I gave a joint public lecture at Carnegie's Root Hall in DC extolling the accomplishments of the last 20 years and giving a glimpse of what was yet to come: GMT, TESS, JWST, and WFIRST, maybe followed by HabEx or ATLAST. Paul Hertz and Charlie Bolden were in the audience. I gave them a shout out and asked them both to stand while the audience applauded their efforts on behalf of NASA's Exoplanet Exploration Program, which was co-sponsoring the event. Root Hall was packed, and the atmosphere was electric. We were moving forward faster than we thought possible, a rare event in the onerous struggles to build flagship-class NASA missions. Warp speed, Mr. Sulu.

Full Stop Ordered: The TMT project announced plans to resume site construction on Mauna Kea in November 2015, prompting protestors to seek an emergency order from the Hawaiian Supreme Court to put the project on hold. The Court accepted their argument and issued a temporary halt. Hawaii's Supreme Court then decided on December 2, 2015, that the TMT construction permit for Mauna Kea was invalid.

The protesters had done their job, and they could go back home, as the Hawaiian courts had taken complete control of the situation. The TMT cost estimate had risen another $100 million to $1.5 billion, perhaps factoring in the time lost trying to placate the Hawaiian protestors and the possible need to find a new site off the Big Island.

The TMT board accepted the Court's ruling and announced that they were assessing the situation. Evidently a new construction permit would be needed. The first permit had required years to obtain, and the protestors were likely to be even better armed in fighting any attempt to secure a second permit.

The fate of the TMT project became unclear at best. Would some of the TMT international partners decide to cut and run, and join either GMT or E-ELT instead? Chile continued to be supportive of ground-based astronomy in the Andean foothills.

Congress Says Go for It!: Congress released the Final Appropriations Bill for FY 2016 on December 16, 2015. Enshrined was language ordering NASA to spend $90 million on WFIRST, just as the Senate had specified. The Mikulski fix was in. All systems were go: WFIRST-AFTA was now just plain WFIRST, and WFIRST was on a fast track toward implementation.

We learned at an ExoPAG meeting in Kissimmee, Florida, on January 3, 2016, that WFIRST had passed a mission design review and was now planned for launch to an L2 orbit, like JWST, rather than an Earth orbit, like HST. This meant that WFIRST's orbit would be compatible with a star shade, as an L2 orbit provided enough real estate for the star shade spacecraft to maneuver all around the WFIRST telescope and hold its position long enough for a WFIRST camera to snap a photo of an exoEarth or two. The WFIRST project had no funds for such a star shade, but that might come later, once Astro 2020 had a chance to consider the situation.

NASA wasted no time in accepting Congress's direction, and the NASA Program Management Council formally approved a new start for WFIRST on February 17, 2016. WFIRST was now in Phase A, exceeding our earlier expectation. Celebratory emails circulated in the exoplanet community spreading the good news. We were on our way.

Dawn of Another New Era: On February 11, 2016, a team of astrophysicists announced the first unambiguous detection of gravitational waves by the Laser Interferometer Gravitational Wave Observatory (LIGO). Another new era in astronomy began on that day. The detection was made on September 15, 2015, when the two elements of LIGO, twin interferometers located thousands of miles apart in the states of Louisiana and Washington, detected the same signal. The time delay between the arrival times of the two signals could be used to triangulate on the source of the waves, given that gravitational waves propagate at the speed of light as predicted by Albert Einstein over a century earlier. The waves had the signature shape of the merger of two black holes, a catastrophic event that occurred 1.3 billion years ago. The observed chirp implied that a binary system composed of two black holes with masses of 29 and 36 solar masses spiraled inward and merged into a single 62-solar-mass black hole, while releasing energy equivalent to 3 solar masses—an unprecedented event indeed.

Astronomers had been searching for gravitational waves for decades, beginning with Joseph Weber's work in the 1960s with large cylinders composed of aluminum (Weber bars). Weber believed that he was able to detect gravitational waves, but the claims could not be reproduced when another scientist built a similar detector. By the mid-1970s, the claims had been dismissed. In 2002, LIGO began operation, with a much more sensitive instrument concept, yet the twin arrays still found no signals after 8 years of operation. LIGO was then shut down for 5 years to undergo an upgrade; and as soon as Advanced LIGO began taking data in the fall of 2015, the first gravitational waves were detected. Plans were underway to improve LIGO even further, and the LISA Pathfinder spacecraft was currently operating in space, preparing the way for a future space gravitational wave observatory. A new field of astronomy has been born, with a bright future indeed.

The parallels between the advent of this new era of gravitational waves and the discovery of the first confirmed exoplanet around a sun-like star are remarkable. Exoplanets had been predicted to exist for centuries, and failed searches had been underway for several decades. In October 1995, Swiss astronomers announced the discovery of an exoplanet with a mass half that of Jupiter in orbit around a solar twin, 51 Pegasus. The total mass of this binary system was about 65 times smaller than that of the pre-merger binary black hole system, and the resulting wobble of 51 Pegasus was just as difficult to detect for the first time. Previous claims for detecting the wobbles of stars caused by orbiting Jupiter-mass planets had all failed—most spectacularly the claims for a Jupiter around Barnard's star, first advanced in the 1960s, and finally debunked by the mid-1970s by an astronomer using a different telescope. The difference was that the claim for the first exoplanet (51 Peg b) was confirmed 1 week later by a team using a different telescope. Paul Butler was largely responsible for this confirmation of the existence of 51 Peg b. Although the LIGO confirmation came just 7 ms later at the second interferometer, Butler had to wait a week for his next scheduled telescope run to see if 51 Peg b was real or not. It was.

Soon after I arrived at DTM in 1981, a student of John Archibald Wheeler, one of the pioneers of general relativity and the study of gravitational waves, asked for a copy of the computer code that I had written for my PhD research on the formation of binary stars. The code calculated the three-dimensional hydrodynamics of gas clouds and could be used to study the spiraling inward and merger process of two neutron stars. I spent a few days inserting comments into my code so that a novice could run it, wrote a 9-track data tape with the code, and sent the package off to the student at the University of Texas. He thankfully acknowledged receipt of the package but then went silent.

A few years later, I asked a Texas colleague about what happened to the student, as I never saw any publications result from the plan to calculate the gravitational waves that would result. I found out that after struggling with my computer code for a few months, the graduate student left the University and took up a new career as an itinerant preacher.

Go for Launch: NASA approval to begin development of WFIRST was in part a result of the final appropriations bill passed on December 16, 2015, by the U.S. Congress. Buried on page 27 of the lengthy document were two budget directives of profound importance: Congress directed NASA to begin development of a robotic mission to orbit and land on Jupiter's icy moon Europa, and to begin development of WFIRST. The terse Congressional budget language did not specify the motivating scientific factor for these two directives, but the implication was clear to those of us in the know: NASA was entering into a race to find the first hard evidence for life beyond Earth. The Europa Clipper mission was given a target launch date of 2022, to be followed shortly thereafter by the launch of WFIRST, NASA's first space telescope intended to search for life on Earth-like worlds beyond the Solar System. Along with the ongoing search for evidence of life on Mars, Congress was doubling down on the bet that life exists beyond Earth and that NASA should be given the expensive, flagship-class missions needed to perform the search.

The View from NASA HQ: The Astrophysics Subcommittee met at NASA HQ on March 15, 2016, where we heard from APD Director Paul Hertz that the Astro 2020 Decadal Survey would need to decide if NASA should remain committed to a 10% share of ESA's third planned large space mission, cleverly named L3, which would detect gravitational waves. Now that LIGO had proven their existence, the thirst for more detections could not be denied. NASA's 10% commitment to the L3 mission carried with it a cost cap of $150 million; but if NASA was going to increase its contribution to L3, then something else in the APD budget would have to be sacrificed. ESA planned the launch of L3 in 2034, so there was time to consider how to proceed; but the fact that L3 was to be an optical laser interferometer on a grand scale meant that the final cost of such a revolutionary space telescope might be equally grand. L3 would have three free-flying spacecraft, separated by millions of kilometers, in order to detect the faint ripples of the space–time continuum, that is, the distance between the spacecraft caused by the passage of a gravitational wave. Nothing quite like L3 had ever flown before, though ESA had launched a technology development mission, LISA Pathfinder, in December 2015, with science operations scheduled to begin in March 2016. LISA Pathfinder would be a key test of whether the technology needed for L3 was ready for prime time.

As the newest child in the astrophysics nursery, L3 was beginning to threaten the lunch of its older siblings. WFIRST was still on schedule for a 2025 launch, though an earlier launch would be possible if extra funds could be made available somehow. In fact, WFIRST was in effect stealing lunch from other projects in the NASA APD by virtue of Congressional control of the NASA budget. The FY16 Congressional budget for APD specified that $76 million would be spent on WFIRST, whereas most of the rest of the APD would suffer a 7% reduction. Hubble and JWST would be spared these draconian cuts, consistent with the continued budgetary clout of Maryland's retiring senior senator, Barbara Mikulski. WFIRST was now estimated to cost somewhere in the range of $2.7 billion to $3.2 billion, depending on the schedule and size of the launch vehicle that would be needed. Hertz would have to spread around the pain of the 7% APD reduction, but WFIRST would not feel even a pinch.

Associate Administrator John Grunsfeld was roaming the halls outside the APS meeting and buttonholed me to talk about his ideas for building a space telescope large enough to image exoEarths. A recent AURA report had proposed building a gigantic 12-m space telescope, twice the size of JWST, that would be able to detect, image, and characterize dozens of Earth-like planets as well as serve as an ambitious successor to both Hubble and JWST. Even that, though, was not large enough to slake Grunsfeld's thirst.

Administrator Charlie Bolden had been the pilot on the Space Shuttle *Discovery* flight in 1990 that carried HST up to orbit the Earth. As a former astronaut who had been literally instrumental in repairing HST on three of the five Hubble servicing missions, Grunsfeld had a passion for keeping the human element involved in astronomical space telescopes. Unlike Hubble, JWST was not designed to be serviceable, so Grunsfeld was thinking big about what astronauts might be able to do to help with the next flagship mission following WFIRST. WFIRST was being planned for robotic, not human, servicing. He suggested to me that exoplanet folks should consider proposing to build a space telescope so large that it would have to be assembled in space by astronauts. He implied that certain other federal agencies with classified projects were already doing similar operations in space.

The 12-m space telescope proposed by the AURA study, the High-Definition Space Telescope (HDST), was in some sense the foundation for one of the four large missions being studied for possible consideration by the Astro 2020 Decadal Survey. The HabEx concept was chosen as one of the four large-mission studies of most relevance to exoplanets, as well as one entitled the Large UltraViolet Optical InfraRed space telescope, or LUVOIR. Grunsfeld suggested that the LUVOIR Science and Technology Definition Team (STDT) should not be limited in their thinking by a measly 12-m segmented telescope, which, like JWST, could be folded up like an origami and launched to space inside the fairing of a single launch vehicle. Why not consider one large enough to require several launches to bring up the pieces, as well as a launch or two to bring up a few astronauts with power socket wrenches and an assembly diagram? Wait, what?

The expectation was that Hillary Clinton would win the 2016 Presidential election, and that the incoming Administration in 2017 would likely continue the strong support provided by the Obama Administration from 2008 to 2016 for Earth science missions at NASA. In order to keep pace with this expected stiff competition, Grunsfeld felt that NASA astronomers needed to think big and discover enough exoEarths to study to ensure that Earth itself was just one of dozens of habitable worlds worthy of consideration.

Agency-Wide Message to All NASA Employees: On April 5, 2016, an email was broadcast to all NASA employees. As a member of the APS, I was considered a "special government employee" for several days a year, and as a result I got stuck on the email list for people who actually work at HQ. The resulting multiple daily emails are essentially spam to me, but every once in awhile something interesting appears in my email inbox. The subject line for this email read "John Grunsfeld Announces Retirement from NASA." That one got my attention. After nearly 40 years at NASA, Grunsfeld was stepping down. Grunsfeld was only 58 years old and in good shape, so why would he step down when things were going well for NASA's SMD?

The rumor mill the next day presented a plausible explanation for this early departure from SMD. Charlie Bolden was expected to step down as NASA Administrator when the new Presidential Administration took office in January 2017, so that job would become available. NASA had never had an Administrator who was promoted from within: they had all been selected from positions outside NASA. The suspicion thus was that Grunsfeld had stepped down well ahead of time so that he could be considered as a strong contender for the next NASA Administrator. Being a former astronaut, like Bolden, needless to say, only strengthened his hopes of winning the top spot at NASA.

Congress Cuts and Caps Costs: The Senate Appropriations Committee on April 21, 2016, approved a funding bill for FY17 that cut funds from other NASA programs in order to bolster the support for the human space flight side of the house.

The APD fared reasonably well, though it was mandated by the Senate to spend $120 million on WFIRST, $30 million more than was requested by the Obama Administration. WFIRST had become an unfunded mandate, as APD did not receive an extra $30 million earmarked for WFIRST. In addition, the Senate decreed that NASA should place a firm cost cap on the total cost of the WFIRST mission of $3.5 billion, and that WFIRST should be launched by 2024. JWST's budget request was untouched, but the other NASA Science Divisions took substantial cuts, especially planetary science, which did not seem to be a priority during the Obama years. The horse race between finding life on planets beyond the Solar System versus on Solar System bodies, such as Mars and Europa, seemed to be tilting toward the APD's derby entrant, WFIRST.

10

Ground-Based Telescopes Score a Hat Trick

What I like about scientists is that they are a team, so that one need not know their names.

—Lord John Wilmot, UK Minister of Supply, 1945–1947

Three Earth-size planets were announced on May 3, 2016, to have been found in orbit around a low-mass star named TRAPPIST-1 by its discoverers, led by Belgian Michael Gillon (see Figure 10.1), evidently an enthusiast for Belgian Trappist beer. Gillon and his Geneva Observatory team had already struck pay dirt back in May 2007 (described in detail in TCU) when he announced the discovery of the first transiting super-Earth, Gliese 436b, albeit a hot super-Earth, with a mass almost 23 times that of Earth and an orbital period of only 2.6 days. Gliese 436b was not a good place to think about looking for life, but it proved the point that super-Earths could be detected by transit photometry, and Gillon had shown that he knew how to do that, even from the ground.

Gillon had begun a new ground-based transit search in 2010 using a 60 cm telescope in Chile named the TRAnsiting Planets and Planetesimals Small Telescope (TRAPPIST). Like Kepler's discoveries, the TRAPPIST-1 planetary system had been discovered by measuring the minute and periodic dimmings of the central star as each of three planets passed in front of the star, temporarily blocking a small portion of the star's light from being seen by the TRAPPIST telescope in Chile at ESO's La Silla Observatory.

Kepler had to be launched into space, outside the Earth's atmosphere, in order to detect the tiny dimmings caused by an Earth-sized planet. So how was TRAPPIST able to do this from a ground-based observatory located only about 1.5 miles above sea level? The reason was that Kepler was designed to be able detect Earth-sized planets orbiting sun-like stars. The Earth is about 100 times smaller in size than the Sun, meaning that at most, the Earth only blocks a fraction of the Sun's light equal to its cross-sectional area, which is given by the Earth's radius, squared, times *pi*

Figure 10.1 Michael Gillon, the leader of the team that discovered the seven transiting exoplanets of the TRAPPIST-1 system (Courtesy of NASA/JPL-Caltech).

(3.1515926536 ...), divided by the area of the Sun. With a radius 100 times smaller, that means that Earth only blocks about 1/10,000th of the Sun's light, a dimming too faint to be seen through the flickering effects of the Earth's atmosphere.

The secret to the success of the TRAPPIST-1 system discovery was that the host star, TRAPPIST-1, was a low-mass, red dwarf star, with a mass of only about 8% that of the Sun. More importantly, this low mass meant that the radius of the host star was only about 12% of that of the Sun. As a result, an Earth-sized planet orbiting TRAPPIST-1 would block out about 1% of the red dwarf star's light, a dimming that can be measured from the ground.

Even more remarkably, these three new exoplanets were orbiting at distances from their faint host star that might allow liquid water to be stable on their surfaces, that is, they were close to being habitable. The TRAPPIST-1 star is less than 1,000 times as bright as our Sun, but the fact that the three new planets orbited their star every few days meant that they were about 100 times closer to their star than Earth is to the Sun. Since the apparent brightness of an object like a star depends on the inverse square of the distance, that meant that TRAPPIST-1 heats these planets just a few times more strongly than the Sun heats the Earth. If there were other unseen planets in this system, orbiting with periods a bit longer than the three found to date, there was a good chance that one of them might turn out to be a bona fide habitable world. Kepler had found a couple of such worlds orbiting lower-mass stars than the Sun, but those discoveries were of systems that were typically hundreds of light-years away. The TRAPPIST-1 system was only about 39 light-years away. Gillon wisely decided to continue to monitor the TRAPPIST-1 system and began to apply for even more telescope time on other ground-based telescopes, such as one of ESO's four VLTs in northern Chile, and on NASA's Spitzer Space Telescope, still in orbit around the Sun on an Earth-trailing orbit, like Kepler.

Meanwhile, about 15 miles north of La Silla, as the condor flies, lies Carnegie's LCO. Both La Silla and LCO share the same access road off the main north–south highway in Chile, Ruta Cinco (Route 5); and each observatory is easily seen from the other observatory. I had begun an astrometric planet search on LCO's 2.5-m (100 inch) du Pont telescope in 2002, and I used to joke with the Swiss astronomers on La Silla, who were searching with their Doppler spectroscope, that we should wave to each other at sunset before we disappeared into our respective domes to begin the night's work.

In 2007, we began using our homebuilt camera, the Carnegie Astrometric Planet Search Camera (CAPSCam), using the same Hawaii-2RG detector array that was being used in some of JWST's cameras. We were following about a hundred low-mass, red dwarf stars, looking for the periodic wobble across the sky that might imply the presence of an unseen, gas-giant-size exoplanet. One of the stars that we were following was an M8 red dwarf we called AJW26 because it was the 26th target star suggested by my DTM colleague Alycia J. Weinberger. We added AJW26 to our observing list in 2011, and observed it about three times a year, looking for the long-period wobble that is most easily found by astrometric detection efforts.

When news of the discovery of TRAPPIST-1 was published in the *Washington Post* on May 3, 2016, I casually went to the journal where the research paper was published, *Nature*, to learn more about this interesting system. One of the first things I did was check the location of the star on the sky, the two coordinates that astronomers refer to as right ascension and declination, and compared those numbers with those in my list of target stars. It became immediately clear that TRAPPIST-1, which was also named 2MASS J23062928-0502285 (a star studied by the 2 Micron All-Sky Survey), was the same as our target star AJW26. Whoa. I immediately dug out my stack of preliminary results for our astrometric planet search to see if there was any hint of a wobble in a red dwarf star that was now known for certain to harbor exoplanets.

We had taken data on AJW26 on 13 separate occasions by that time, enough to reveal any large-amplitude, long-period astrometric wobble, if there was one; but there was no indication from our preliminary data reduction of anything that might be a real signal. I put the Gillon et al. *Nature* article in the stack of data printouts and added a note to the observing list that this one had at least three planets. That meant it was well worth keeping on our target list. Who knew what else might be lurking in the outer regions around this faint red dwarf?

Carnegie Evening Gossip Page: The Carnegie Institution throws a grand party for its scientists and invited guests in the spring of each year, along with an hour-long lecture by a distinguished, often Carnegie, scientist. The annual event is held in Carnegie's administrative headquarters building in Washington, DC, a few blocks north of the White House, at the corner of 16th Street and P Street. P Street, as it is affectionately known within the Institution, holds a large auditorium, ballroom, and marbled rotunda that provide a perfect venue for this annual celebration of science. In fact, the venue is so magnificent that the building is frequently rented out for weddings, with the happy couple then often placing a half-page photo spread of the event in the *Washington Post.* Oddly enough, the Center for the Study of Responsive Law, Ralph Nader's creation, rents its office space in the basement of P Street, an arrangement that allowed Ralph to creep up a side staircase and enjoy the raw oyster bar on Carnegie Evenings.

The Carnegie Institution was founded in 1902 with an initial $10 million endowment from Andrew Carnegie, who had sold Carnegie Steel and retired from business in 1901 in order to dedicate the remainder of his long life to philanthropy. Carnegie worked with President Theodore Roosevelt to create something new in Washington, DC, an institution that would be dedicated to basic scientific research rather than yet another university. The story is that Carnegie thought that tycoon Leland Stanford had made a terrible mistake by founding Leland Stanford Junior University, named after Stanford's son, so close to the UCB campus, as the inevitable competition between these neighboring universities would be detrimental to both. Although that perception has since been proven to be spectacularly misguided, the fact that Carnegie chose to create a new type of institution, dedicated to pure research, was a decision that was celebrated annually at Carnegie Evening.

The 2016 Carnegie Evening was held on May 5. My DTM colleague, Alycia Weinberger, was the featured lecturer, and she presented a wide-ranging talk about her own and other astronomers' studies of the protoplanetary disks observed around young stars, where planetary systems form, and of the debris disks left over once the planet formation process had run its course. As a result of this noted speaker and her topic, as well as the general intention to not miss a spectacular party with an open bar, astronomers were out in full force that evening, including a number of our colleagues from the Carnegie Observatories, located in Pasadena, California, as well as some of those colleagues who had taken leave of absence from the Observatories to work on planning and designing the GMT. The new Director of the Observatories, John Mulchaey, was in attendance as well, as Carnegie Evening is usually scheduled to coincide with a meeting of the Carnegie Board of Trustees, where the Directors learn the latest plans for the Institution.

The topic of the ongoing fundraising battle between the competing TMT and GMT projects arose in my conversation with Mulchaey, with the added severe complication for the TMT effort being the recent decision by the Hawaiian Supreme Court that the TMT's construction permit for Mauna Kea was invalid. It was clear that the TMT folks needed to be thinking about observatory locations other than Mauna Kea. John confided in me that yes, indeed, the TMT folks were quietly considering other mountaintops and had even contacted him to inquire about Carnegie's Las Campanas mountain in Chile, where GMT would be built.

That was a stunning revelation. Some of us had long wondered why the TMT and GMT camps had not figured out a way to combine their projects. When asked whether there were likely to be enough institutional commitments and private donations to be able to fund the construction of *two* giant telescopes, the cynical reply was that there was not enough money to build even *one* of them. Still, the two combined projects would seem to have a better chance of succeeding than either one would have on their own. After all, TMT partner Caltech and the Carnegie Observatories (then named the Mount Wilson Observatory) had run the Palomar Observatory in Southern California for decades. Palomar is the home of the Hale 200-inch (5.1 m) telescope, the world's largest for 45 years, named after George Ellery Hale, who founded both the Mount Wilson Observatory and the Palomar Observatory. However, the marriage between Caltech and Carnegie was dissolved when Carnegie decided to shift its resources away from Palomar and to develop the LCO in the Southern Hemisphere in the early 1970s. Although some bad blood remained from that mid-1970s divorce, most of the characters from that era were no longer active at either institution, and so relations had warmed again—but not necessarily to the extent to seriously consider combining GMT and TMT in some uncertain fashion. Still, the idea was intriguing to consider. How long would it take for the TMT team to decide to find a mountaintop that would not provoke perpetual legal battles?

Extrapolating into the Unknown: On May 9, 2016, the NASA ExEP Chief Technologist, Nick Siegler, sent out an email to a few dozen astronomers who were engaged in the search for exoplanets. Nick attached a copy of a new paper just submitted to the ApJ by Wes Traub, the ExEP Chief Scientist. Traub had extended his previous analyses to include the first 17 quarters of Kepler observations. Kepler had found a few Earth-like planets by this time, but it was just beginning to scratch the surface of the most interesting portion of exoplanet discovery space, that is, exoplanets with the same size and stellar heating as our Earth. Traub had come up with a remarkable, even stunning result: the Kepler data seemed to be implying that G dwarf stars, that is, stars like our Sun, each have on average about *five* planets. Considering that our own Solar System, which we know in great detail, has only eight planets (sorry, Pluto), the fact that Traub estimated that the typical Sun-like star has at least five relatively short-period planets, with orbital periods less than 2 years, meant that our Solar System, with only four such planets (Mercury, Venus, Earth, Mars), was not as crowded as the inner region around a typical G dwarf star.

Of those five planets per star, on average, each G dwarf was estimated to have about *one* Earth-like planet. Traub was thus claiming that η_E, the single number that was the raison d'être for the Kepler Mission, was unity. While the Kepler Mission team was unwilling to make such an estimate based on their data, preferring to wait until the end of the mission, Traub was quite willing to take a stab at estimating η_E. Traub was heavily involved in planning for future NASA space telescopes that would attempt to image Earth-like planets; and in order to balance the immense cost of such a mission with the expected return on the investment, that is, the mission yield of exoEarths, one needed to have a good estimate of η_E. If Kepler was not ready to estimate η_E, Traub was ready to take the leap.

The Kepler team held a press conference on May 10, 2016, heralding their latest results concerning the exoplanet yield. Although Kepler had found several thousand likely exoplanets, it was known that a good fraction of these might turn out to be imposters, for a variety of reasons. The Kepler team had been putting a significant amount of effort into performing this process of validating their candidates, which meant using a fair amount of ground-based telescope time to examine the stars suspected of hosting transiting planets. The most valuable test was to use Doppler spectroscopy to see if the star was indeed wobbling in its radial velocity with the same period as the inferred Kepler transit orbital period. Besides providing a convincing check—seeing the same planet with two very different detection methods—a firm Doppler detection yielded the mass of the exoplanet, which, when combined with the planet's radius from the transit depth, yields the density of the planet. Planetary density was the gold standard for differentiating between low-density, gas giants and high-density, rock and iron terrestrial planets.

However, Doppler spectroscopy was only possible for the brightest Kepler candidate stars, so other techniques were used as well to weed out the interlopers, such as background eclipsing binary stars. In order to achieve the photometric precision needed to detect the 1-part-in-10,000 dimming caused by Earth-like planets, Kepler's optics were purposely defocused a bit in order to spread out the star's light onto a "postage stamp," typically a square of 5 × 5 pixels on the 95-megapixel CCD detector that was the heart of the Kepler telescope. This slight blurring meant that there was a finite chance of other background stars also being included in the postage stamp that was recorded and sent to NASA Ames for analysis by the Kepler team. If one of those background stars happened to be an eclipsing binary—that is, a binary star system where the orbit of the binary is aligned so that one star periodically passes in front of the other star—then a distant, faint, eclipsing binary could make the coupled light from the brighter, foreground, Kepler target star and the distant, background, binary star appear to be coming from a single foreground star that happened to have a small, transiting planet. That would be a false positive, a binary star pretending to be an exoplanet, and the Kepler team needed to eliminate these imposters.

One means to uncover these deceptive systems was through detailed, pixel-by-pixel examination of every pixel in the postage stamp, searching for hints that the perceived overall dimming was arising from a single pixel or two where the eclipsing binary was hiding. This approach worked for binaries that were separated from the target star by a few pixels, but it failed if the eclipsing binary was very close to the line of sight to the target star. In that case, Kepler team members would apply for time on ground-based, moderately sized optical telescopes with adaptive optics imaging systems, systems designed to minimize the blurring caused by the Earth's atmosphere, allowing in some cases an otherwise unknown binary star to be discovered right next to the Kepler target star. If these high-resolution images revealed an eclipsing binary, the planet candidate would be ruled out. However, one of the most powerful validation techniques was searching for more than one planetary companion in the Kepler light curves: if there were several planets dimming the same star by different amounts, and with different orbital periods, something that a single background eclipsing binary could not imitate, then there was an excellent chance that all of those dimmings were caused by planets. Kepler had found that many stars had more than one planet, so this became a key validation method.

At the May 10, 2016, press conference, Kepler Project Scientist Natalie Batalha presented the latest summary of the status of the 4,302 planet candidates found in the DR24 catalog, which had been released in 2015, and which was also the basis of Traub's estimate of η_E. Of this total, 984 had already been confirmed to be true planets by one or the other of the validation techniques, and Batalha was happy to announce that the latest follow-up work had now confirmed another 1,284 planets. Of the remaining candidates, 707 were known or suspected to be false positives and so could be discarded. However, there were still another 1,327 that could not be confirmed, but the signs were that they were planets as well. That meant that a total of 3,595 of the 4,302 candidates were known or likely to be exoplanets: Kepler had a 84% success rate of finding true exoplanets.

Batalha noted that the final Kepler catalog was required to be released by October 2017. Perhaps then we would learn what the Kepler team thought was the best estimate of η_E. Until then, we would have to rely on Traub's estimate of η_E about equal to 1, an easy number to remember. If Traub was even close to being correct, potentially habitable worlds were as common as stars in the night sky—one would not be able to look up at night without seeing numerous stars with rocky planets where some other life forms might be looking back at us. Smile, and wave hello.

Congress Cracks the Whip: The top-ranked large mission in the 2011 Planetary Science Decadal Survey (Planet 2011) was MAX-C, a robot intended to explore the Martian surface, looking for evidence of either fossil life or at least hints of the prebiotic chemistry thought to lead to life in a suitable planetary environment. MAX-C was planned to "collect, document, and package samples for future collection and return to Earth," as long as it could all be done for no more than about $2.5 billion. The clever, but obvious, implicit rationale for collecting these precious samples, but leaving them on Mars, was that once we knew that we had found samples that might prove that life had once existed on Mars, if only they could be thoroughly examined in terrestrial laboratories, astrobiologists and NASA would move heaven and Earth, if necessary, to secure the federal funds needed to send another robot that would pick up the cached samples and return them to Earth for analysis. This second effort would be considerably more expensive than calling up Uber Eats on your smart phone and waiting for your meal to arrive—the second robot would have to not only make it to Mars, land successfully, and retrieve the cache, but it would have to take off from the Martian surface, find its way back to Earth, and deliver the samples safely to the NASA team waiting below: without unintentionally contaminating the samples with Earth microbes, or the Earth with Martian microbes. The federally mandated environmental impact report for such a sample return mission might make the regulatory and political obstacles to building, say, a new nuclear power plant, seem minor in comparison.

But there was a second-place large mission recommended in Planet 2011: the Jupiter Europa Orbiter (JEO), which was intended to "explore Europa to investigate its habitability." Europa is one of the four Galilean satellites of Jupiter, with the other three being the innermost Io, and then the outermost Ganymede and Callisto. These four were named for their discovery by Galileo Galilei in 1610, and they constituted the first direct, visual proof that celestial bodies could be seen to orbit around a more massive body, just as Copernicus hypothesized was the case for the Solar System's planets orbiting around the Sun.

NASA's robotic spacecraft reconnaissance of Jupiter and its extensive satellite system, beginning with the launches of Voyager 1 and 2 in 1977, followed by the Galileo Mission, and continuing to this day with the Juno Mission, currently in orbit around Jupiter, had shown that Europa was an icy world slightly smaller than the Moon. With a silicate rock mantle and possibly an iron–nickel core, Europa resembles the bulk physical properties of the Earth, but with an important difference. The NASA robotic images of Europa revealed the surface to be a chaotic ice sheet, broken, sheared, and striated into fascinating patterns of cracks. The smoothness of Europa's surface implies that these ice sheets might be floating on top of an ocean of liquid water. Coupled with Europa's oxygen atmosphere, the idea that life might exist inside Europa has fired the imagination of planetary scientists and astrobiologists for decades.

While planetary scientists debate the thickness of the icy crust, the idea that deep down below, strange life forms might be eking out their existence became a prime motivation for the Planet 2011 second-place ranking of the JEO concept. Perhaps evidence of life could be seen in the cracks, where subsurface water might erupt? Plumes of water vapor may have been seen on Europa; and if a spacecraft could pass through a plume and grab a sample, it might be possible to determine if subsurface microbes exist, without having to land and drill down through an icy crust of uncertain depth.

Even members of Congress were interested in the possibility of life on Europa, which had so far taken a back seat to the hunt for signs of life on Mars within NASA. Representative John Culberson of Texas, who succeeded Frank Wolf as chair of the House Appropriations committee in control of NASA's budget, had single-handedly been inserting funds into the NASA budget for several fiscal years earmarked for use by JPL in planning a Europa mission. Culberson was pushing not just for the Europa orbiter called for by Planet 2011 but for an orbiter and a lander that would sample the environment of one of the icy cracks, and perhaps catch a sniff of the water

Plate 1. Artist's depiction of the NASA Kepler space telescope, launched in March 2009 (Courtesy of NASA).

Plate 2. Artist's depiction of the NASA Spitzer space telescope, launched in 2003 (Courtesy of NASA/JPL-Caltech).

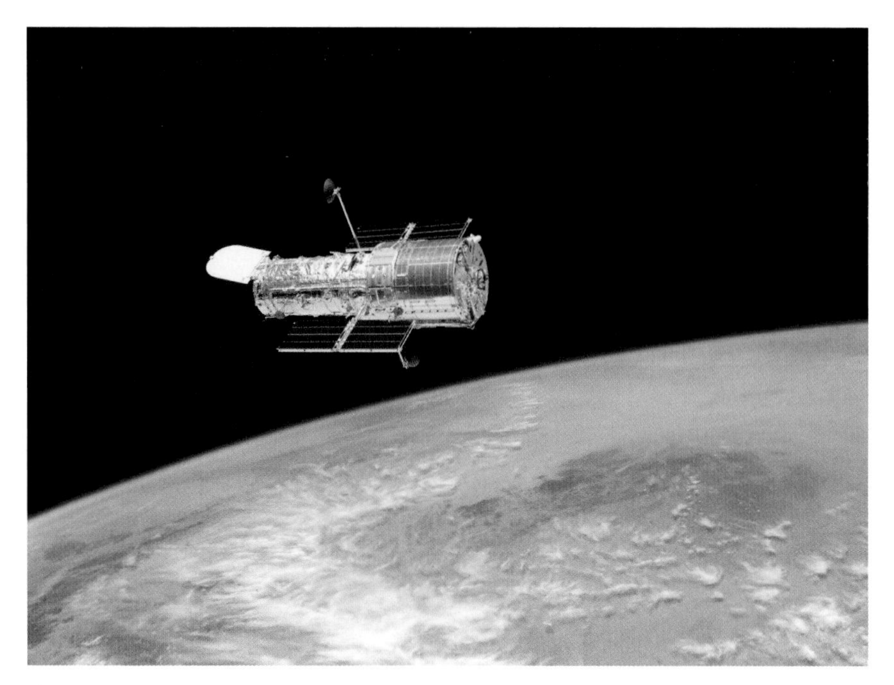

Plate 3. NASA Space Shuttle photograph of the Hubble Space Telescope upon completion of the final servicing mission in May 2009 (Courtesy of NASA/STScI).

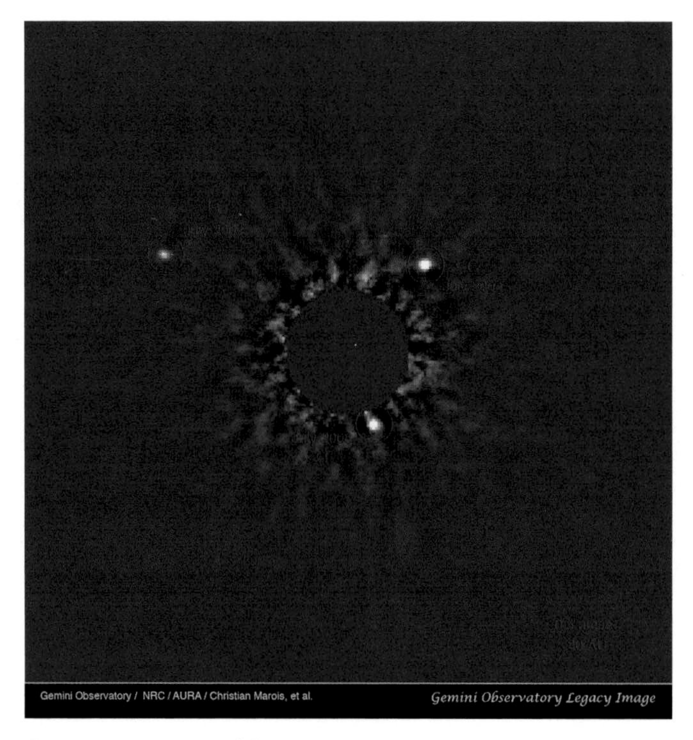

Plate 4. Adaptive optics image of the HR 8799 system, showing the first three exoplanets discovered (b, c, d) by the U.S. Gemini North telescope (Courtesy of Gemini Observatory/NRC/AURA/Christan Marois, et al.).

Plate 5. Artist's conception of the GJ 581 planetary system (Courtesy of NSF).

Plate 6. Artist's conception of the Gl 667 triple star system exoplanets (Courtesy of ESO).

Plate 7. Artist's conception of the seven-planet TRAPPIST-1 system (Courtesy of NASA/JPL-Caltech).

Plate 8. The Alpha Centauri triple star system, consisting of Alpha Centauri A (left), Alpha Centauri B (right), and Proxima Centauri (faint star inside red circle), the closest stars to the Solar System (Courtesy of Skatebiker at English Wikipedia).

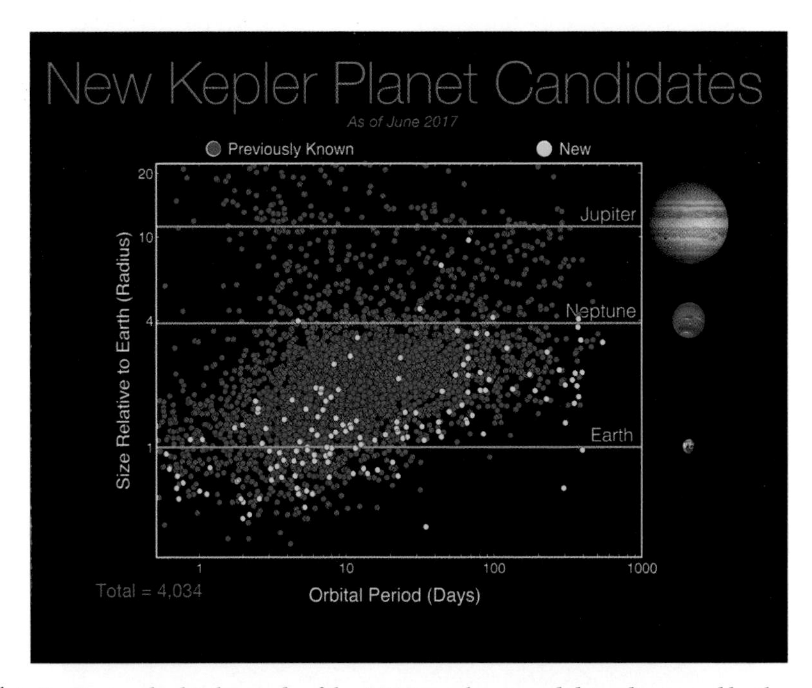

Plate 9. Sizes and orbital periods of the 4,034 exoplanet candidates discovered by the Kepler Mission as of June 2017, from the final DR25 data release (Courtesy of NASA).

Plate 10. The ESA Gaia space astrometry telescope, launched in December 2013 (Courtesy of ESA).

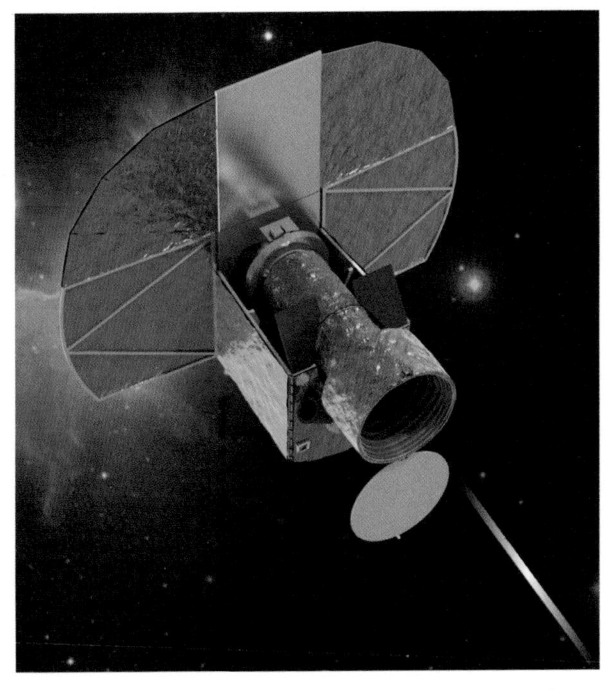

Plate 11. Artist's depiction of ESA's CHEOPS space transit telescope (Courtesy of ESA).

Plate 12. Artist's depiction of NASA's TESS, the Transiting Exoplanet Survey Satellite (Courtesy of NASA/GSFC).

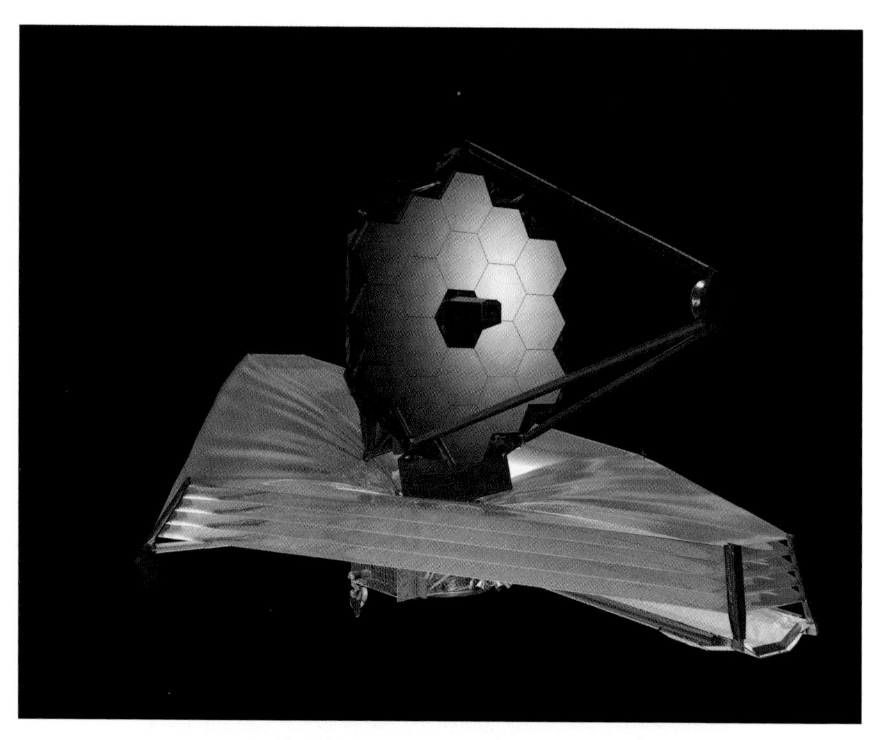

Plate 13. Artist's depiction of NASA's James Webb Space Telescope (JWST), showing the segmented telescope on top of its multilayered thermal sunshield (Courtesy of NASA).

Plate 14. Artist's depiction of the NASA WFIRST space telescope (Courtesy of NASA).

Plate 15. Artist's depiction of the Giant Magellan Telescope (GMT) on Cerro Las Campanas, Chile (Courtesy of GMT).

Plate 15. Artist's depiction of the TMT, the Thirty Meter Telescope, on Mauna Kea (Courtesy of the TMT International Observatory).

Plate 16. Artist's depiction of the E-ELT, the European Extremely Large Telescope, on Cerro Armazones, Chile (Courtesy of ESO/L. Calcada).

vapor emanating from the ocean down below. The JPL mission planners were all too happy to figure out how to build and fly such a complicated mission: that is what they do, and they do it exceedingly well.

It would be expensive, but the attraction of finding definitive proof of life beyond Earth was enough to compel Culberson to insert another $260 million for the Europa combination mission in the FY 2017 House appropriations bill released the week of May 19, 2016. Added to the $395 million already directed by Culberson to this mission study in the previous 4 FYs, this was a sizeable down payment on the Europa mission concept. Culberson requested that NASA PSD present a 5-year funding profile for the mission in the FY18 budget request. A Congressman was cracking the whip and telling NASA to get with the program, that is, with Congressman Culberson's program.

Culberson was supportive not only of the Europa mission but also of the search for life outside the Solar System. The same FY 2017 appropriations bill that called for a large increase in funds for planning a Europa mission also called for having the ExEP spend $10 million in 2017 "to develop the technology for a possible Star Shade demonstration mission using the Wide-Field Infrared Survey Telescope." Clearly, the folks at JPL, where the ExEP is based, or somewhere else, had Representative Culberson's ear. A star shade for WFIRST could be a marvelous means for allowing this relatively small (2.4 m) telescope to perform major league science discoveries. The 2015 Exoplanet Star Shade (Exo-S) Rendezvous study of what a 34-m-diameter star shade could accomplish in combination with a 2.4-m space telescope suggested that one or two habitable, Earth-size planets would be detected by searching up to 50 nearby stars. This estimate was based on assuming that η_E was equal to a measly 0.16, or 16%. If Wes Traub was correct, and η_E was closer to 100%, the expected yield from a star shade operating in tandem with WFIRST might be six times higher, and as many as a dozen exoEarths could then be detected and their atmospheres given at least a first look for possible signs of biosignature molecules.

Culberson was placing his bets on both red and black—when the roulette wheel finished spinning, the ball was likely to land on either finding life inside or outside the Solar System: that is, unless life on Earth was unique, or if life was rare in the cosmos. That would be the equivalent of the roulette wheel ball ending up in one of the two green pockets instead of one of the 18 red or 18 black pockets—Culberson, and we, would lose. The roulette wheel design suggested that Culberson's chances of winning were good: the ball should only end up in a green pocket about 5% of the time. One does not turn down a bet like that, unless you know the game is rigged. Would Mother Nature rig the wheel on us?

Another battle was shaping up, with Culberson's help. Within the NASA PSD, the Martian life folks were battling for supremacy with the resurgent Europeans, whereas just down the hall on the third floor at NASA HQ, the PSD was in a friendly competition with APD's plans for searching for signs of life on nearby exoEarths. Who would win? The two Division Directors placed a bet on their own Division winning the race. Or might ground-based telescopes steal the thunder of the claim of the first evidence for life beyond Earth?

The FY 2017 appropriations bill also explicitly stated that the "priorities outlined in the decadal surveys for . . . Planetary Science . . . [and] Astrophysics . . . shall drive NASA mission priorities." The Decadal Surveys are now, and ever shall be, the Bible for NASA SMD decision makers: follow the commandments laid out there, or perish.

11

And That's Not All

Space-travel is utter bilge.

—Sir Richard van der Riet Woolley, The Astronomer-Royal, 1956–1971

Buried on page 60 of the FY 2017 House appropriations bill was an even more astounding requirement for NASA to achieve beyond the Europa orbiter and lander mission: NASA was directed to begin studies and development of the technology needed to be able to accelerate a spacecraft to one-tenth the speed of light, or about 30,000 km per second (18,600 miles per second). That may not seem like an impossible task to undertake, but it must be understood in the context of the highest speeds ever achieved by NASA's spacecraft. The New Horizons mission to Pluto, launched in 2006, was the fastest spacecraft ever launched by NASA, with enough speed to ensure that it would leave the Sun's gravitational pull and head off to interstellar space once it had passed by Pluto and completed its main mission. New Horizons was launched with an unprecedented rocket configuration: an Atlas V first stage with five booster rockets, an Atlas Centaur second stage, and an Alliant Techsystems Inc. (ATK) Star third stage. Even still, New Horizons was on its way to the outer Solar System with a top speed of about 16 km per second (10 miles per second). The Voyager 1 spacecraft managed to achieve an even higher final speed out of the Solar System, but only as a result of the gravity assists produced by close encounters with both Jupiter and Saturn, the so-called gravitational slingshot trick. As a result, Voyager 1 is currently leaving the Sun at a speed of about 17 km per second (11 miles per second). Evidently the high speeds achieved by both of these pioneering missions leave something to be desired, compared to the language in the FY 2017 House appropriations bill: they are both about a factor of 1,800 times too slow.

What was the motivation for handing NASA this difficult task? The bill made it clear: NASA should plan on achieving this goal by "the one-hundredth anniversary, 2069, of the Apollo 11 Moon landing." The launch target would be the closest star system to Earth, the Alpha Centauri binary-star system, lying a tantalizingly close 1.3 parsecs (4.4 light-years) away. Alpha Centauri A is a sun-like star with a mass only 10% greater than that of our Sun, while Alpha Centauri B is also a sun-like star with a mass 10% smaller. They orbit around each other on an eccentric orbit, with distances varying from about 11 times the Earth–Sun distance (roughly Saturn's orbital distance from the Sun) to as far as about 36 times the Earth–Sun distance (roughly Pluto's orbit). With an orbital period of about 80 years, the Alpha Centauri binary system is far enough apart that Earth-like, habitable planets could orbit stably around either or both stars. Claims had been made that Doppler spectroscopy had found evidence for an exoEarth around Alpha Centauri B, but it was a hot exoEarth, much too close to its star to allow water to be stable. But if there was one Earth-size orbiting Alpha Centauri B, Kepler had taught us that there was a good chance that there were more rocky planets there as well.

If a spacecraft could be launched and accelerated to a speed of 0.1 the speed of light, that would mean it would only take about 44 years to reach Alpha Centauri, and if the spacecraft could snap a few photos of any planets in the Alpha Centauri binary star system, and send us all an interstellar Instagram, that photo stream would arrive back on Earth about 48.4 years after the original spacecraft was sent off on its way. Although 48.4 years may seem like an impossibly long time to wait, the Washington National Cathedral required 83 years to construct, and most people would agree that the wait was worth it. The same could be said of the visionary goal of returning close-up photos of what may be the closest exoplanetary system.

With the *New Horizons* launch technology, such a mission would require about 79,000 years to reach Alpha Centauri, plus 4.4 more years for the photos to arrive at Earth. Let's just round that off to an even 80,000 years or so, shall we? Such a time interval would try the patience of even the most saintly of the exoplanet aficionados. NASA might even find evidence of life on Mars or Europa before then, which would tend to spoil the Alpha Centauri excitement a bit, and hand the life-finder prize to the PSD rather than to the APD.

If NASA could figure out a way to accelerate a spacecraft to that speed, clearly that would be a game-changer for interstellar flight. Although the appropriations bill directed NASA to put together a detailed plan, a roadmap for how to proceed, there were no new funds added to accomplish this effort, a classic example of an unfunded mandate. The roadmap was due to be submitted to the House in 2017 and the roadmap was to include an estimate of the funding requirements; so there was an implicit promise of some dedicated funding if the roadmap looked promising.

The roadmap was to consider a number of speculative approaches to achieving a speed that was a significant fraction of the speed of light, including nuclear fusion (proposed in 1992 by hydrogen fusion bomb creator Edward Teller), matter–antimatter annihilation reactions, solar wind sails, and directed beams of electromagnetic (light) energy. While these approaches are thoroughly imaginative, it is instructive to consider the constraint applied by Einstein's theory of special relativity regarding the energy needed to accelerate any object to a significant fraction of the speed of light. The applicable relativistic formula shows that in order to accelerate an object to 0.1 the speed of light, the amount of energy required is about 0.5% of the object's rest mass energy, which is equal to the object's mass times the speed of light, squared. That small fraction of an object's rest mass energy may not sound like much at all, but remember that the speed of light is a big number, and squaring it compounds immensely the problem of the energy needed. For example, the rest mass energy of a 1-ton automobile is about equal to the energy contained in a trillion gallons of gasoline. Reducing that amount of gas by the factor of 0.005 results in a total fuel requirement of about 5 billion gallons, or at least $10 billion at a cost of $2 per gallon. Automakers have yet to figure out how to build a 1-ton car with a fuel tank capable of holding 5 billion gallons: 5 billion gallons is about equal to the amount of petroleum consumed every 5 days by the United States. That is a lot of gas.

Hey, I Had That Idea First: While the FY 2017 House bill was released the week of May 19, 2016, the idea about achieving speeds close to the speed of light had already been tossed around by Russian Internet billionaire Yuri Milner and his Breakthrough Starshot team. The previous month of April 2016, Milner's team made the announcement that they had settled on a concept that just might work, even if Albert Einstein was correct—which, of course, he was. Rather than try to launch an interstellar probe to Alpha Centauri, along with all the rocket fuel needed to accelerate it to close to the speed of light, the Breakthrough Starshot initiative proposed to launch a cluster of tiny, 1-g, fuel-less spacecraft, each with a thin sail that would be blasted with light from a bank of powerful lasers back on Earth and thereby accelerated to high speeds. Considering that no fuel needed to be carried, and that the rest mass energy of a 1-g spacecraft is a million times smaller than that of the automobile previously considered, Milner needed to use lasers to deliver the equivalent of a mere 5,000 gallons of gas to the 1-g spacecraft; clearly this would be a bargain with gas at $2 per gallon.

Although the $10,000 cost of gas for a single, 1-g microsatellite seemed reasonable, the fact that lasers would be needed to provide the electromagnetic momentum that would accelerate the microsatellites meant that an immense bank of high-powered lasers would need to be built and operated from the Earth for an extended time period. Estimates of the cost of the entire operation began with an opening bid of, what shall we say, how about $10 billion? That would be pricey even for Milner, who saw his $100 million investment as seed money for something that might catch the attention of other deep-pocketed investors. The first $100 million would go a long way toward thinking about whether the scheme might work, or whether it should just be dropped altogether. Milner even hinted that he was looking into contributing to efforts to use existing ground-based telescopes to take an image of the putative Alpha Centauri planets, without having to wait until the next decade for the extremely large, ground-based telescopes to become operational and do the job, or the many decades later for one of his microsatellites to send a snapshot home.

E-ELT Doubles Down on Its Bet: On May 25, 2016, the Europeans signed a $448 million contract for the construction of the dome and support structure for the E-ELT, the 39-m-diameter gigantic telescope planned for the Cerro Armazones mountain in the coastal range of the Atacama Desert in northern Chile. The press release noted that the E-ELT would have a light-collecting area greater than that of all of the optical telescopes on the Earth, at the Filene's Basement price of about $1.2 billion.

The E-ELT would be located at an elevation of about 3,046 m (9,993 feet) at the summit of Cerro Armazones, slightly higher than the plan for the 24.5-m GMT, with an elevation of about 2,500 m (8,200 feet) on Cerro Las Campanas, about 250 miles farther south in the Atacama Desert. Construction of the E-ELT site was scheduled to begin in 2017 and anticipated its first light in 2024. The European commitment to the race to build the first extremely large telescope was now inked and legally binding, with $448 million on the table. Would the TMT and GMT call, or raise, the billion-dollar bet?

Star Shade Compatibility Study: On June 7, 2016, the NASA APD announced formally that the WFIRST mission should be planned so that it would be compatible with a possible star shade intended for imaging and characterizing spectroscopically exoplanets beyond those that would be studied with the WFIRST coronagraphic instrument (CGI), which was already baselined as an instrument during the ongoing Phase A study. This meant that there had to be a radio receiver/transponder on the WFIRST spacecraft that would allow the WFIRST telescope to measure the distance to and position of the distant star shade, as the two spacecraft performed a distant minuet that would require a precise alignment between the center of the star shade and the center of the optical axis of the telescope. A star shade would allow fainter exoplanets to be imaged than what the CGI could accomplish, and it would not be as limited in terms of seeing more distant planets, that is, those with orbits more than a few times the Earth–Sun distance. While any star shade would have to be approved by the upcoming Astro 2020 Decadal Survey, it made perfect sense to see if star shade compatibility could be maintained in the WFIRST design, provided that the cost, schedule, and risk increments associated with this added capability were modest.

WFIRST would be the only major, flagship-class mission for the next decade, or longer, so if we missed this opportunity to do first-class exoplanet science, the penalty would likely be another decade of waiting. On the other hand, when there is only one major mission on the drawing table, something called the Christmas tree effect comes into play: all the interested parties want to add a few more baubles here and there in order to make sure that this mission will be able to do all sorts of things besides its primary goal—and, of course, to provide the extra features that would keep the engineers at various NASA centers busy for a few more years. HST was the classic example of the Christmas tree effect, though its splendid performance over its decades in orbit argued in favor of this effect rather than against it. JWST was the next holiday tree to become adorned with something for everyone, though its rapidly mounting cost and schedule problems had nearly forced its cancellation back in 2011. The WFIRST team and APD were both were keenly aware of the mortal dangers of the Christmas tree effect, and so the star shade compatibility study had better make sure that the added costs were indeed incremental, if not insignificant, if the star shade option had any hope of surviving the years before the Astro 2020 Decadal Survey gave it a closer look.

You Can See Mexico from Here: The NASA ExoPAG held its 14th meeting on the eastern shore of San Diego Bay, a location demanded by our desire to hold our relatively small meeting of exoplanet enthusiasts immediately prior to the much larger, spring meeting of the AAS in the San Diego Convention Center. The AAS meeting could be counted on to attract several thousand astronomers to San Diego, and we hoped that a few would arrive a day or two earlier to attend the ExoPAG meeting and find out what was happening in the world of NASA's plans for exoplanet exploration.

We learned on June 11, 2016, that the "final occurrence rate products" for the Kepler Mission, that is, the final tally of how many planets of what size, and of what orbital period, had been found in the 4 years that Kepler was fully operational, were scheduled to be released in April 2017, less than a year away. We were getting agonizingly close to learning the Kepler estimate of η_E. Would it even be released in April 2017, or would we have to wait for Kepler's final catalog to appear in October 2017? Once released to the world, would it agree with the optimistic estimate of η_E of essentially unity proposed by Wes Traub? It was hard to imagine it being much higher than unity, but an estimate much less than unity would mean that we were not as close to finding nearby habitable worlds as we hoped we were. Alpha Centauri was still the closest star system, but did it host a habitable world?

We also learned at the time that the launch of TESS would likely be delayed from August 2017 to December 2017, a delay not caused by any problem with building the telescope or its spacecraft but with getting a ride up to orbit. TESS would be hitching a short ride on a SpaceX Falcon 9 rocket to its highly unusual, highly elliptical orbit around the Earth and would then spend several years using transit photometry, like Kepler, to look for potentially habitable worlds. TESS would examine 500,000 of the brightest, closest stars to Earth, located over the entire sky, rather than stare in a single direction, as Kepler had done. Given the relatively short TESS prime mission lifetime of 2 years, that meant that the orbit of the typical transiting planet would have to be measured in months, rather than years; and hence, if a TESS planet candidate was going to be in the habitable zone, it would have to be much closer to its star than Earth is to the Sun. That meant that the star would have to be considerably less bright than the Sun or else the only water to be found would be in the form of scaldingly hot steam. That in turn meant that most of TESS's habitable worlds would be in orbit around low-mass, red dwarf stars. Such M dwarf stars are commonplace in our galaxy, and TESS would have no shortage of target stars to follow: the 1,000 closest red dwarf stars would be on the top of the TESS observing list.

The only problem was that SpaceX had suffered a disaster with a Falcon 9 rocket mission to the International Space Station in June 2015, with a second stage that exploded in flight a few minutes after it was launched from Cape Canaveral. SpaceX soon discovered the source of the problem and successfully returned to flight, but given its aggressive launch schedule, and the payloads waiting in line ahead of TESS, this meant that SpaceX would not be able to send TESS off and on its way until a few months later than was originally planned. TESS was intended in large part to find nearby habitable, transiting worlds for follow-up observations by JWST; and given that JWST was still planned for launch in October 2018, essentially by Congressional decree, even a few months, delay in launching TESS would cut significantly into the time that TESS would need to discover the desired habitable exoplanets for follow-up by JWST.

Waiting for the Clouds to Part a Hemisphere Away: One nice aspect of remote observing is that if the weather turns bad at the observatory, instead of being trapped in an observatory dome on a remote mountaintop with only one's laptop for amusement, an idle astronomer can wander around one's own home and do whatever one would do for amusement on a normal evening, such as have a leisurely reading of the *Washington Post*'s print edition, rather than being limited to squinting at the online content. On the night of June 19, 2016, I took advantage of the rare cloudy night at LCO in Chile to finally watch on cable television the film "Interstellar," which had premiered back in November 2014. I had not bothered to try to see the movie in the intervening 2 years, in spite of its box office success and in spite of the fact that I had spent a memorable day at Caltech in 2006 with Steven Spielberg and Kip Thorne discussing ideas for a movie concept that eventually metamorphosed into "Interstellar" a decade later.

Watching "Interstellar" did create an unusual personal juxtaposition. That evening I was actively continuing my search for outer gas giant planets in orbit around nearby red dwarf stars, hoping to find a few such exoplanets, which might hint at the possibility of inner, shorter-period, rocky exoplanets that might be habitable, based on the example of our own Solar System. This was real work on real stars in the real universe. It was at times a boring, but always painstaking effort, lasting for over a decade so far, full of promise, yet with nothing concrete in the sense of a new exoplanet to show for our efforts to date. The movie "Interstellar" had leaped completely over the true difficulties of searching for new worlds, much less traveling to them and back, and instead presented the adventure as nothing more involved than an inexplicable, phantasmagoric ride through a wormhole, a ride worthy of an Orlando theme park. My life was definitely not imitating that sort of art.

Kip Thorne had worked as a scientific consultant on "Interstellar" and advised the digital engineers who created the images of what the region around a black hole might look like should one be so unfortunate as to be within its grasp. But

Thorne had real science of his own to contemplate in June 2016, as it had just been announced at the San Diego AAS meeting that LIGO had detected a second gravitational wave late on Christmas Day in December 2015. This Christmas gift consisted of the merger of two somewhat smaller black holes than had first been detected merging in September 2015: the inferred masses were 14 and 7.5 times that of the Sun, compared to 36 and 29 solar masses in the original detection. The announcement paralleled in a way what had happened when the first confirmable detection of an exoplanet was presented in October 1995 by Swiss astronomer Michel Mayor: a few months later, in 1996, it was announced at an AAS meeting that Geoff Marcy and my DTM colleague Paul Butler had evidence for two more exoplanets. In both cases, it was clear that a new field of astronomy had been born.

Thorne was what we graduate students in the Physics Department at UCSB in the mid-1970s called a "Four-Star General," one of the main players in the field of developing Einstein's general theory of relativity. Thorne was the "T" in MTW: the abbreviation for Charles W. Misner, Kip S. Thorne, and John Archibald Wheeler's massive, 1,279-page tome entitled, simply enough, *Gravitation.*

MTW appeared in 1973 as I was entering graduate school at UCSB, at the same time that the English translation of Victor S. Safronov's pioneering 1969 Russian book *Evolution of the Protoplanetary Cloud and Formation of the Earth and the Planets* appeared in print. Safronov needed a mere 206 pages to explain how the Solar System formed from a cloud of gas and dust, a remarkable study in brevity compared to MTW. Both volumes were far ahead of their times: it would be over 20 years before the first definitive extrasolar planet was discovered in 1995, and another 20 years before the first gravitational wave was seen in 2015.

I bought both books, and read much of MTW while taking a class in General Relativity from another Four-Star General, UCSB's James B. Hartle; but I read and worked through every single page of Safronov's treatise, re-deriving each equation therein in preparation for developing a PhD thesis project on the subject of planetary system formation with my advisor, UCSB's Stanton J. Peale. But what interested me the most about Safronov's development was the assumption in the first chapter that the planetary system was formed from a flattened, rotating disk of gas and dust in orbit around the newly formed Sun. This hypothesis dated back to the thoughts of French philosopher Pierre Simon de Laplace in the late 1700s, when the first observations of what would turn out to be spiral galaxies were confused for planetary systems in formation. Given these initial conditions for the gas and dust disk, termed the solar nebula, Safronov proceeded to assemble a plausible sequence of events for forming progressively larger dust agglomerates, planetesimals, planetary embryos, and finally, fully formed planets like the Earth. Although a reasonable assumption, by 1975, we had no clear observational evidence for the existence of such protoplanetary disks around young single stars. In fact, the state of the art in 1975 of our understanding of the star formation process itself was primitive at best. We did not know how in the world single, binary, or multiple stars formed, though ideas were beginning to emerge about how rotating gas clouds might collapse and fragment into binary and multiple protostars. I dropped Safronov's book and instead began a PhD thesis project at UCSB devoted to developing the theory of star formation: I figured that once that basic problem had been solved, we would know better if Safronov's assumptions were reasonable or not.

12

Proxima Centauri b Arrives on Stage

Work, Finish, Publish.

—Michael Faraday, 1791–1867

Paul Butler agreed to give a seminar for DTM's weekly department-wide seminar series on July 7, 2016. As seminar czar for the 2015–2016 academic year, I was looking forward to introducing Paul for his talk and to the cessation of my chores as the czar at the end of the month. In my introduction, I noted that I had gotten to know Paul for the first time at an exoplanets meeting in Lisbon, Portugal, in 1998, the same year that my first popular book on exoplanets, LFE, was published. Paul was one of the stars of that book, and I had been astonished to learn in Lisbon that he was looking to leave his position in Australia for a more permanent position. As soon as I returned home from Lisbon, I began a campaign to hire Paul at DTM, in spite of the fact that DTM already had 15 staff members at the time and no empty positions to fill. Buttressed by the excitement of the birth of the new field of science that was presented in LFE, I won my case, and Paul moved to DTM in 1999, firmly establishing DTM as a player in the new world of exoplanets.

Paul's seminar talk was modestly entitled "Surveying nearby stars for planets." It turned out that he had a bombshell to deliver, though he cautioned the audience that the new result was formally under embargo, as it was contained in a paper to be published soon by *Nature*. The paper in question had been accepted just 3 days earlier, and as one of the 31 co-authors on the paper, Paul felt free to spill the beans to the DTM audience.

The new paper dealt with a search for a possible Earth-mass planet on a HZ orbit around the closest star of all to the Sun, a faint red dwarf star called Proxima Centauri, or Prox Cen for short. Prox Cen is the often-neglected third member of the Alpha Centauri system, which is dominated in mass by the solar-mass binary stars Alpha Cen A and Alpha Cen B, the latter of which appears to have a hot, Earth-mass planet. Prox Cen is sometimes called Alpha Centauri C in honor of its close association with the Alpha Centauri binary star system, making this a triple-star system. Prox Cen has a mass of only about 12% that of the Sun, barely qualifying it as a star. Objects with masses less than about 7.5% of the Sun's mass are unable to shine for billions of years by fusing hydrogen nuclei into helium nuclei, and so are termed brown dwarfs rather than stars. Prox Cen is lightweight, but it is massive enough to avoid that ignominious designation and fate.

Prox Cen is slightly closer to Earth at 4.25 light-years (1.30 parsecs) than either Alpha Cen A or B, both lying about 4.37 light-years (1.34 parsecs) away. If you had to pick a nearby star to have a habitable planet, you would want to pick the closest one of all, and that was Prox Cen. There are over 10 times as many red dwarf stars in the solar neighborhood as there are G dwarf stars like the Sun, so the chances of the nearest star being a red dwarf are actually quite good.

In spite of being the closest star to the Earth, humans cannot see Prox Cen with the naked eye, as it is over 10,000 times fainter than the Sun at the wavelengths of light used by our eyes. It was not even discovered by astronomers until 1915. As a red dwarf star, Prox Cen emits most of its light at longer, redder wavelengths than visual wavelengths, so that its total luminosity, added up over all wavelengths of light, is only about 600 times less than that of the Sun. Alpha Cen A and B are easily visible to the naked eye, though they are so close together that the human eye cannot tell that they are a pair of stars, rather than a single, very bright star. In fact, Alpha Centauri (A and B combined) is the third brightest star in the sky, with only Sirius (the brightest of them all, also a binary) and Canopus being brighter. Like two-wage-earner families in a materialistic society, binary stars tend to outshine their single-star competition.

Prox Cen had been first suspected of having an exoplanet in 2013, based on an analysis of archival Doppler spectroscopy by Mikko Tuomi, an astronomer who specializes in analyzing Doppler data in the search for hidden exoplanet signals. The observations were taken on the 3.6-m telescope at ESO's La Silla Observatory and on one of the four 8-m VLT telescopes at ESO's Paranal Observatory and then stored on a publicly available archive. Both ESO observatories are located in Chile: Prox Cen, like Alpha Centauri, lies so far south, 28 degrees from the southern celestial pole, that it can best be seen only in the Southern Hemisphere. While there seemed to be a Doppler wobble indicative of a planetary mass companion in the archival data analyzed by Tuomi, with a short period of about 11 days, the problem was that Prox Cen is an active star, like many M dwarf stars. That is, the star's spectral lines, used for the Doppler shift analysis, can dance around on time periods of days to weeks to months, as a result of star spots and flares on the star's rotating surface. Such stellar activity can masquerade as a Doppler wobble caused by a real exoplanet, so caution was in order. Still, there seemed to be something inducing a signal with a period of about 11 days.

Given this situation, it was clear that more data needed to be taken, and a lot of data at that, including simultaneous monitoring of Prox Cen's stellar activity in order to try to disentangle the Doppler variations caused by the star's own activity from that caused by any short-period exoplanet. A major effort was needed to garner the large amount of dedicated telescope time that would be required.

Guillem Anglada-Escude, now of the Queen Mary College of the University of London, took the initiative and launched the effort to understand definitively what was going on with Prox Cen. Dubbed the "Pale Red Dot" (PRD) initiative, Guillem assembled a large team of primarily European astronomers with access to Southern Hemisphere telescopes. HARPS was used to take Doppler measurements of Prox Cen on almost every night between January 19 and March 31, 2016, for a total of 54 new epochs of HARPS data. Combined with the archive of 90 previous epochs of HARPS data, and 72 epochs from the VLT spectrometer UVES, Anglada and his team could now perform an exhaustive search for true Doppler signals indicative of an exoplanet. In addition, the PRD team included simultaneous photometric monitoring of Prox Cen as measuring whether or not a star is brightening and dimming with a certain period is an excellent means for deciding if that periodicity is intrinsic to the star, or extrinsic, that is, caused by an exoplanet. If the photometric monitoring happened to detect the signal of a transiting planet, that would be a spectacular added bonus, as combined with a limit on the planet's mass from the Doppler signal, the transit data would constrain the true mass of the planet, as well as its size—and so the mean density of the unseen world would be determined.

While the PRD campaign did not detect a transiting planet, the Doppler data implied that Prox Cen was indeed orbited by an exoplanet every 11.2 days, as had been suspected originally by Tuomi, now relegated to the 30th place on the roster of 31 co-authors of the *Nature* paper as a result of the listing of the co-authors' names in alphabetical order. But the big news was that this exoplanet appeared to have a mass as low as 1.3 times that of Earth. Not only that, but the exoEarth orbited Prox Cen at a distance 20 times closer than the Earth is to the Sun; and given Prox Cen's faintness, that meant that Prox Cen b, the official astronomical name for the planet, was likely to be the right temperature to be able to sustain liquid water on its surface.

Let's see now, what do we have: an Earth-mass planet with Earth-like stellar illumination. Bingo! We had all hit the jackpot. Mother Nature had been kind to us, giving us a possibly habitable, Earth-like planet orbiting the closest star of all of the hundreds of billions of stars in the Milky Way galaxy. This was a secret that could now be revealed by Paul, in spite of the formal *Nature* embargo, to the chosen few in attendance at the DTM seminar that day in July.

Once the press release about Prox Cen b was out, the Breakthrough Starshot crowd would have a new, even better target than Alpha Centauri A and B for launching their microsatellites. Prox Cen had leapt to the top of the target list, and it could not possibly ever be displaced, as there is little chance that there are any closer stars that have not yet been discovered, much less one with a potentially habitable Earth 2.0.

Paul was an active participant in the PRD initiative from the start, and I became aware of the effort at an early phase as well, as a result of my close relationship with Guillem and my extensive background in the world of exoplanets. Guillem had been a postdoctoral associate with Alycia Weinberger and me at DTM from 2008 to 2011; and during those fruitful years, Guillem had written the data analysis pipeline used for our CAPSCam astrometric planet search on the 2.4-m du Pont telescope at LCO. Guillem had been an extremely valuable and talented colleague during his postdoc; and Paul, Alycia, and I were quite sad to see him leave after a little over 3 years with us. We would have loved to be able to hire Guillem as a staff member, but once again there were 15 on the DTM staff, and this time we were unable to convince our Director to hire another astronomy group member. Our hiring of Paul in 1999 was ultimately painless for DTM only because one of the older DTM astronomers agreed to move out to California and join the Carnegie Observatories in Pasadena, thereby restoring the magic number of 15 DTM staff members. This time the oldest astronomy group member was me, and I was not quite ready to leave DTM, even if it could be guaranteed that Guillem would get my DTM staff position, which of course would never be guaranteed in the first place. Carnegie staff positions are few and precious, and the normal replacement process for a staff member involves a year-long, worldwide search for the best candidates.

Guillem had invited me to write the opening commentary for the launch of the PRD campaign. It was released online on January 14, 2016, entitled "Pale Blue Dot, Pale Red Dot, Pale Green Dot, . . ." playing on Carl Sagan's famous description of the Earth when seen from afar as a "pale blue dot," and on Guillem's focus on the "pale red dot" that is Prox Cen. Given the red light emitted by Prox Cen, any Prox Cen planet with a reflecting atmosphere might also appear to be a pale red dot. The "pale green dot" was intended to highlight the ultimate goal of the worldwide exoplanet race: a planet with evidence of life, such as the green terrestrial foliage associated with photosynthesis and chlorophyll, and the last three dots ". . ." implied that the discoveries would continue, even once we found a pale green dot.

Six months later, Guillem's PRD campaign had indeed shown that Prox Cen had a habitable planet. The discovery finally hit the streets on August 24, 2016, timed for the publication of the *Nature* paper. The question immediately became, how are we going to take a picture of Prox Cen b? Could we do it from the ground with existing large telescopes, or would we have to wait for the next generation of ground-based telescopes? Or would a future space telescope get the coveted photo first? The race to image Prox Cen b was about to start. Game on.

The Senate Ups the Ante: At the summer meeting of the Astrophysics Subcommittee at NASA HQ on July 20, 2016, we learned that the Senate raised the bet on NASA's future space exoplanet telescope plans. While the House Appropriations Subcommittee had added $10 million for a star shade for WFIRST, the Senate Appropriations Subcommittee had added an extra $30 million for WFIRST, on top of the $90 million in the President's FY 2017 request. The House–Senate conference committee would have the final decision; but when both houses of Congress are seeking to raise the spending levels for WFIRST and a possible star shade, and both seek to maintain the funding level for JWST, things could not be imagined to go much better, given the overall constraints on the federal budget and its ongoing, seemingly perpetual deficit problem. Presumably, the stunning Kepler hints about the frequency of Earth-like planets, coupled with the TRAPPIST-1 and Prox Cen b discoveries, helped to prime the pump for this Congressional largesse.

How About the Canary Islands?: With the possibility of building the TMT in a timely manner on the intended site of Mauna Kea becoming less and less likely, the TMT partners were considering their options. The TMT project hoped that their legal problems in Hawaii would be settled in their favor by early 2017, but clearly they needed to look for a backup plan in case that did not happen. The TMT project decided that they needed to start construction on the observatory site, wherever it might be, by no later than April of 2018, a date less than 2 years away at that point, and that time schedule required a building permit for Mauna Kea to be issued by early 2017. The TMT was now estimated to cost about $1.4 billion. The cost could only climb if there were any further delays in beginning construction on a mountain somewhere. Pouring concrete also has a way to make a dream as grand as a giant telescope seem like something that might actually happen, which could encourage additional visionary partners to join the TMT consortium and help provide the remaining funds needed to complete the observatory.

Mauna Kea was the prime location in part because then the TMT would have much of the northern half of the sky to itself—both the E-ELT and the GMT were being built in northern Chile, well south of the equator, by about 24.5 and 29 degrees, respectively. It would be a shame for the international astronomical community if all three of the next generation of giant telescopes were built in the Southern Hemisphere. In addition, there would be a major negative effect on the future of Hawaiian astronomy if the Second Battle of Mauna Kea was lost: how could anyone plan on building another major new astronomical facility on the summit of this sacred mountain after two straight defeats?

The First Battle of Mauna Kea had been lost in 2006 when Hawaiians successfully halted plans to add six, 1.8-m "outrigger" telescopes to the Keck Observatory on Mauna Kea. These outriggers were to have been coupled together to form an optical wavelength interferometer that would be able to discover Neptune-mass exoplanets by following the wobble of their central stars as the star and planet orbited the center of mass of their system. Using the Keck telescopes for astrometric planet detections was one of the major reasons for NASA's investment of one-sixth of the total cost of the twin 10 m telescopes of the Keck Observatory (as described previously in LFE). Astrometric planet detection was a favored planet detection method prior to the discovery of 51 Peg b in 1995 by the Doppler spectroscopy method. However, the HIRES spectrograph on the Keck telescopes turned out to be well worth NASA's $35 million investment in the Keck Observatory in terms of discovering and characterizing exoplanets, from the era of 51 Peg b through to the era of Kepler, K2, and the upcoming TESS mission.

In spite of the Keck outriggers' loss of the First Battle of Mauna Kea, the TMT project was now fully engaged in the Second Battle of Mauna Kea, though the battle was not going well. If Hawaii was not a realistic possibility for the TMT, the search was on for other possible sites in the Northern Hemisphere. However, other Northern Hemisphere sites had already been considered early in the TMT development process, and it had been decided then that Mauna Kea was the best site in the Northern Hemisphere, largely because of its location in the middle of the Pacific Ocean, which paradoxically leads to low water vapor, and to great stability in the atmosphere overhead (the seeing), as the oceanic trade winds blow smoothly across the summit of the volcano. Given this situation, the TMT project decided to widen the renewed search to include both hemispheres, evidently including the private overture to John Mulchaey about the possibility of joining TMT with the GMT project in Chile in some way.

A serious contender for Mauna Kea in the Northern Hemisphere had appeared: the major observatory site on La Palma, in the Canary Islands. The Roque de los Muchachos Observatory on the island of La Palma was already the home of the 10.4-m Grand Canaries Telescope, as well as several smaller telescopes. The seeing at La Palma had been measured to be superb, second only to that of Mauna Kea in the Northern Hemisphere. The Canary Island astronomers would welcome the addition of the TMT telescope to their observatory, which would thrust La Palma into the forefront of ground-based astronomical research. While the TMT was reconsidering other Northern Hemisphere sites, it was clear that La Palma appeared to be the best alternative to Mauna Kea.

Given that Spain is a member of the E-ELT project in Chile, one might wonder why Spain would permit the TMT competition to gain a foothold on the Canaries. Perhaps Spanish astronomers involved with the E-ELT were just being altruistic with regard to the desire of every astronomer for all-sky coverage by giant telescopes. But another reason emerges as well: the Canary Islands are an autonomous community of Spain, giving the La Palma astronomers sufficient independence from the mainland to woo the TMT. It was looking like a good time to place a bet on La Palma as the ultimate site for the TMT project to break ground.

NASA Emails a Dear Colleague Letter: By August 2016, the NASA APD folks were eagerly anticipating and preparing for the upcoming Astro 2020 Decadal Survey, slated to begin some time in 2019 or so. Astro 2020 would define the big-ticket items for the APD for the coming decade, and it was important to get an early start on preparing for this critical planning exercise, to be run, as usual, by the NRC of the National Academy of Sciences. The previous NRC survey report had sorted its recommendations for NASA into three categories, based on the expected costs: small-scale activities (typically no more than about $10 million per year), medium-scale activities (total cost of no more than $200 million), and large-scale activities (total costs typically over $1 billion). These three cost categories left a gaping hole in the range of total costs between $200 million and over $1 billion, the latter being the flagship mission class.

While PI-led missions costing less than $180 million could be developed as part of the existing NASA Explorer Program, there was no way for a PI-led mission that blew the cost caps on the Explorer Program to have a chance to compete and win a slot as a NASA-funded mission. Given the steady advances of astronomical discoveries, there were fewer and fewer compelling new mission concepts that could fly on an Explorer Program budget and more and more concepts that could do something really exciting but would cost significantly more. The spectacular success of the $640 million Kepler Mission was a sterling example of a scientific advance that could not have been accomplished on an Explorer budget, though the upcoming TESS Mission is an obvious counterexample, made compelling by the prior success of Kepler, of what could be done with a Medium-Class Explorers (MIDEX) mission budget of no more than $180 million.

NASA uses the "Dear Colleague" letter format to alert the astrophysics community about news items such as upcoming opportunities to propose to perform a service for NASA. In the days when such invitations were delivered by the U.S. Postal Service, rather than the Internet, the letters were printed on paper with the red, white, and blue NASA meatball logo printed in the corner and a salutation that read "Dear Colleague." In the modern email incarnation, the formal salutation is dropped, but the "Dear Colleague" phrase is still included in the subject line of the email, as a reassuring link to the hoary past and a suggestion that NASA really values your collegiality, so please do not label this as spam without reading it first.

The email sent out on August 1, 2016, alerted the recipients that the NASA APD would be attempting to fill this large hole in the NRC costing categories by requesting serious concept studies of missions that would have a total life cycle cost (Phases A through F, with Phase F meaning the closeout of the mission) in the range of approximately $400 million to $1 billion, nicely closing the NRC costing gap. These medium-class missions were to be known as Astrophysical Probes, in spite of the unfortunate semantic association with the purportedly kidnapped guests of alien spacecraft in the American Southwest. NASA would support as many as eight Probe mission concept studies that could fit into that cost range. After a year and a half of consideration, the chosen eight teams would each write a White Paper, a formal report, that would then be submitted to the NRC for consideration by the Astro 2020 Decadal Survey.

The reasoning behind the release of this Dear Colleague letter had become evident in an earlier letter, released on January 14, 2016, which Paul Hertz used to charge the three APD program analysis groups, the ExoPAG, the PhysPAG (Physics of the Cosmos), and the COPAG (Cosmic Origins), with considering two different approaches to handling the question of probe-class missions. Should NASA HQ issue a solicitation for proposals for a few teams to perform studies of specific probe-class mission concepts, with modest budgets (about $100,000 each) for support, or should HQ do nothing and let the community work on developing probe-class concepts all on their own? The three PAGs were charged with consulting their communities and writing a two-page report, to be discussed at the next APS meeting, on March 15–16, 2016. The ExoPAG leapt into action, and we discussed our response to the charge over telephone conferences and email exchanges, delivering our five-page report on time. We finessed the explicit two-page limitation by writing a two-page report with a three-page appendix, with the appendix listing a few of the details of three specific possible probe-class mission concepts of considerable interest to the exoplanet community: first, a star shade to be flown in concert with WFIRST; second, a 1.5-m space telescope to be devoted to spectral studies of transiting exoplanets, from the ultraviolet to the infrared; and third, a 1.2-m space telescope that would perform astrometric planet detections, possibly to as small as an Earth-mass planet orbiting the closest stars. Any one of these three concepts could lead to significant advances, and the ExoPAG's Executive Committee enthusiastically approved our report.

As chair of the ExoPAG, I coordinated our report with those being prepared by the COPAG and the PhysPAG, with the goal being to find common ground, if possible, as speaking with a unified voice was certain to lend more weight to our conclusions than what would likely be the case if all three PAGs reached differing answers to Hertz's charge. Common ground was easy to find, as all three PAGs agreed that it was better to have NASA provide some minimal support for at least a few teams to form and give serious consideration to the science that could be accomplished for $1 billion. The alternative would be a free-for-all, with teams funded by relatively deep-pocketed entities, such as the NASA field centers, having the advantage over those from universities and other institutions that would not be able to provide significant support.

Even when astronomers give their time freely, which is often the case, putting together a serious mission concept study usually involves travel to numerous meetings, along with the cost of the hotels, meals, and a lot of hot coffee. The only concern raised by the joint PAG response to the probe charge was that planning on granting only about $100,000 for each study would not go far in paying engineers to do the studies on Cost, Risk, and Technical Evaluation (CATE) that would help validate the assertion that this mission concept could be flown for no more than about $1 billion and that the technical risks associated with developing, building, and flying the spacecraft were not intolerable. As the Astro 2010 Decadal Survey had learned, the CATE process, performed by the Aerospace Corporation, often reached different conclusions than those reached by the in-house evaluations offered by the mission concept teams. Unlike astronomers, engineers do not work for free. And $100,000 for each probe-class concept study would not buy much in the way of a thorough, reliable CATE.

The APS held a vote on March 15, 2016, on the response of the three PAGs to Paul Hertz's probe-mission charge and unsurprisingly approved the recommendation to proceed with the first option; the Dear Colleague letter sent around on August 1, 2016, was the concrete evidence that probe-class mission concepts would be proposed and considered by the Astro 2020 Decadal Survey. The Astro 2010 total cost gap was being closed.

13

Speaking of the Decadal Survey

Before I came here I was confused about this subject. Having listened to your lecture I am still confused. But on a higher level.

—Enrico Fermi, 1901–1954

On August 15, 2016, the prepublication version of the much-anticipated Midterm Assessment of the Astro 2010 Decadal Survey was released. Given that by definition Decadal Surveys only happen every 10 years, and given the rapid pace of astrophysical research, it was recognized that there was a need to review the progress made toward achieving the goals laid out by Astro 2010 somewhere in the middle of the decade of the 2010s. Astro 2010 recommended that such a midterm assessment be undertaken, and in fact went much farther than that by recommending the creation of the DSIAC, pronounced as "dizzy-ack," an acronym that stood for either the Decadal Survey Implementation Advisory Committee or the Decadal Survey Independent Advice Committee, depending on which pages of the incompletely edited version were consulted. Besides the mid-decade review, Astro 2010 recommended that the DSIAC "should be charged to produce annual reports to the agencies, the Office of Management and Budget, and the Office of Science and Technology Policy" so that NASA could be continually monitored for adherence to the Decadal Survey. The DSIAC reports would go not only to NASA and NSF themselves but also to the OMB and the OSTP, with the implicit threat that if the agencies did not respond to the DSIAC's annual dose of advice, the DSIAC would ratchet the matter up to the next level. As one could imagine, this level of annual management by an external group of well-meaning but inevitably officious astronomers, albeit a group chartered by the august NAS/NRC, stuck in the craw of the agency administrators, who devote their lives to trying to run their agencies in the best possible manner, subject to factors not under their control, such as federal budget constraints, Congressional oversight, the OMB, the OSTP, and the Decadal Surveys: did they really need to have the DSIAC continually looking over their shoulders as a standing committee? Wasn't that what the Astrophysics Subcommittee was supposed to be doing, through the NASA Advisory Council under the Federal Advisory Committee Act?

The NASA APS has been briefed about the results of the Astro 2010 exercise on September 16, 2010, by Roger Blandford, the chair of the Decadal Survey Committee. The Director of the APD at that time was Jon Morse, who also summarized the recommendations of Astro 2010 in his presentation about the state of the APD, including a single line noting the recommendation about the DSIAC, without, however, noting whether the APD had accepted that particular recommendation. The APS held a meeting on February 16 and 17, 2011, when the DSIAC recommendation was discussed at length. As APS chair, I crafted the request that, "given the myriad of often overlapping Congressional and federal agency advisory committees and NRC panels and committees," the APS should invite the chairs of all of these committees to attend the next APS meeting for a discussion, so "that the roles of the APS and the other current standing committees with relation to a potential DSIAC become clearer." I passed the request up the advisory chain of command when I addressed the NAC's Science Committee on March 4, 2011. Compared to the more urgent demands on the attention of the NAC Science Committee, namely, the blossoming JWST debacle, this APS request was not deemed an action item. The next APS meeting was held on July 13, 2011, and the agenda contained no such DSIAC discussion. Jon Morse's presentation that day was careful to list exactly how the APD was responding to the recommendations of Astro 2010, to the extent that the budget realities allowed APD to do so, but made no mention of the DSIAC. By the time of the next APS meeting in October 2011, Jon Morse had been replaced by Geoff Yoder as Acting Director of the APD, as a result of the JWST shake-up within SMD, and the DSIAC had disappeared off all of our radar screens.

Whereas the DSIAC recommendation had evidently been quietly discarded, the call for a mid-decade review was not. In fact, the published Midterm Assessment document stated in a footnote that the mid-decadal committee was carrying out the mid-decadal assessment that had been intended to be performed by the DSIAC, and later dryly noted that the call for a DSIAC to provide annual reviews "was not fulfilled." Indeed, no, it was not. The Midterm Assessment hinted that the Astro 2020 survey was likely to return to this sore point again.

The real hot potato for the Midterm Assessment committee to consider was the status of WFIRST, the top-priority large mission recommended by Astro 2010 for launch in the 2010s. However, given the fact that JWST had not yet been launched, and would not be launched until 2018, there was no way that NASA would be able to launch WFIRST by 2020: large missions typically require at least 7 years to pass through the Phase A-B-C-D NASA mission development process. At best, WFIRST would be ready to go in 2024 or 2025, and that meant that WFIRST was likely to be the only large mission that NASA would be able to launch in the 2020s. This alone suggested that perhaps it would make sense to skip Astro 2020, at least from the point of view of large missions to be launched in the 2020s, and wait for Astro 2030.

However, given the lengthy time period needed to develop a flagship mission, Astro 2020 would still be necessary simply to make recommendations about which large mission should be started in the mid-2020s, after WFIRST was launched, in order for that anointed winner to be ready to go in the 2030s. But first, the question of NASA's plans for WFIRST needed to be evaluated.

On September 16, 2016, the APS was briefed by the chair of the Midterm Assessment committee, Jacqueline Hewitt, a professor of physics at MIT who specializes in cosmological studies. The Midterm Assessment concluded that NASA's adoption of the NRO's 2.4-m space optics for WFIRST combined with the internal coronagraph would result in a significant advance over the science program envisioned by Astro 2010, where the telescope was suggested to be a mere 1.5 m in diameter and would not have the ability to directly image exoplanets that the coronagraph enabled. Not only could the cosmological observations and microlensing surveys proposed for WFIRST be accomplished better with a telescope with a collecting area 2.6 times larger, but the addition of the coronagraph would allow WFIRST to gather data on the atmospheric composition of the exoplanets it would image. Even the Midterm Assessment committee thought that was far out, man.

So far, so good, for the exoplanet aficionados. The Midterm Assessment went on to state carefully that Astro 2010 had only recommended technology development in the 2010s in preparation for a direct-imaging exoplanet mission sometime "beyond 2020" and did not recommend flying a coronagraph on WFIRST, the next flagship train to leave the station. Astro 2010 had decided that the microlensing survey embedded in WFIRST should be sufficient to slake the thirst of the exoplanet community for another decade, while planning for direct imaging some time later on. Still, the Midterm Assessment concluded that given the "remarkable developments in exoplanet discovery" and the rapid improvements in coronagraphic technology that had occurred since Astro 2010, adding the coronagraph to WFIRST was an "appropriate shift in emphasis in WFIRST's science program." Yes, indeed it was, and the recognition of this shift made total sense and served to validate the reason why there was a need to have a Midterm Assessment: things change on time scales much less than a decade in contemporary, cutting-edge astronomy.

However, there is always a "however" about reports like the Midterm Assessment. The Assessment noted that the coronagraph "remains a schedule, cost, and technical risk for WFIRST." Clearly, the costs for WFIRST were climbing. While Astro 2010 had bookkept WFIRST as a $1.6 billion flagship, the 2015 WFIRST DRM study had resulted in an estimated total cost in the range of $2.3 billion to $2.5 billion, in part due to the addition of the coronagraph. WFIRST was now in Phase A, and at the Phase A Key Decision Point (KDP-A), its cost had risen by another $550 million, prompting the Assessment to fret about "NASA program balance," that is, having one large mission's costs grow to the detriment of all the other endeavors in which the APD was engaged. Given that that was exactly what had just happened with JWST's cost growth, the Midterm Assessment was applying the lessons learned from JWST to the case of WFIRST, where it was not necessarily too late to do anything about it. As a result, the Assessment recommended that before WFIRST could pass KDP-B, that is, enter into the next phase of mission planning and development, NASA "should commission an independent technical, management, and cost assessment" of WFIRST, including an assessment of what adding the coronagraph would cost. If this review should result in a WFIRST mission cost estimate that unbalanced the APD portfolio recommended by Astro 2010, then NASA "should descope the mission to restore the scientific priorities and program balance by reducing the mission cost."

There, the "d-word" was finally out, in black and white on the printed page of the Midterm Assessment: *descope*. The more experienced, and therefore cynical, of us involved in NASA's decades-long attempts at developing an exoplanet imaging mission had nervously chuckled when the coronagraph was added onto WFIRST in 2013 that the coronagraph was likely to be viewed by non-exoplanet folks as the top candidate for a descope option, once WFIRST ran into a fiscal brick wall, as it inevitably would, based on recent experience. Such is the possible fate of an instrument that was added on without the explicit blessing of Astro 2010, though the Midterm Assessment committee seemed to accept the coronagraph on the basis of its own merits.

Finally, the Midterm Assessment recommended that NASA rethink its plan to participate with Europe in the L3 mission concept of a space-based gravitational wave detector. Astro 2010 had listed LISA as the third-highest priority for a large mission, but NASA APD simply was unable to afford its estimated $1.5 billion share of the joint ESA mission; and so NASA had dropped down to being a minor partner in ESA's L3 mission concept. Given the loss of the NASA funds, ESA had been forced to descope LISA somewhat for their L3 concept, eLISA. The literally Earth-shaking news of the detection of gravitational waves by LIGO had changed the situation again since the days of Astro 2010: the Midterm Assessment recommended that NASA rejoin ESA in developing an L3 mission concept that was as capable as that envisioned by the original LISA concept. That meant more funds for gravitational wave technology development during the 2010s, as even LISA was not imagined by Astro 2010 to be ready to be launched until 2025. ESA was booking flights in 2034 for its eLISA concept, so there was plenty of time to figure out how best to achieve the goal.

That recommendation in turn raised the question of where the technology development funds needed to ramp up gravitational wave detector research would be found. The likely first place to look would be in the technology development budget line of APD, where Astro 2010 had recommended two medium-scale science themes: exoplanet imaging first, and cosmological inflation second. Given that the Midterm Assessment had noted that the WFIRST coronagraph was actively addressing, at least in part, the first recommendation, it appeared gravitational wave technologists might end up gnawing on any carcasses left behind in the exoplanet technology cafeteria, such as star shade studies. This fear was amplified by a finding that did not make it into the Summary section but was buried back on page 80 of the Midterm Assessment: Finding 4-11 stated that APD was on track to spend more money on exoplanet technology development during the 2010s than had been prescribed by Astro 2010, where a figure in the range of $100 million to $200 million had been recommended. The implication was clear and explicitly stated: the Midterm Assessment committee viewed any growth in exoplanet technology development as being "lower in priority" than gravitational wave technology development. Ouch.

We on the ExoPAG Executive Committee quickly fashioned a response to the Midterm Assessment, and by early October I had submitted our thoughts to Paul Hertz, with a copy to the chair, Professor Hewitt. Although we thanked the committee for doing a fine job overall, the main nit that we wished to pick was the seemingly existential threat to the exoplanet technology development program. We made the point in our response that if the coronagraph should be descoped from WFIRST, and if the remainder of the exoplanet technology development program, the highest-priority, medium-scale activity in Astro 2010, should be reduced or eliminated, the overall effect would be catastrophic. Such an outcome would mean a failure to achieve the highest-priority, medium-scale recommendation of Astro 2010 and would likely result in the delay of the search to detect and characterize habitable worlds by yet another decade. Paul Hertz thanked us for our input; Professor Hewitt did not respond.

Not My Problem: The Midterm Assessment was also charged with reviewing the progress made by NSF on the ground-based astronomy recommendations of Astro 2010, which had given a third-place ranking in the large-scale category to the GSMT, as NSF referred to the competing designs for TMT and GMT. Astro 2010 had recommended that the NSF pursue a 25% share to be invested in the winner of an "immediate partner down-select," that is, in either TMT or GMT. The total cost of the GSMT was estimated to be in the range of $1.1 billion to $1.4 billion in 2010, implying a cost to the NSF in the range of $257 million to $350 million, seemingly a bargain for a large share of one of the next generation of ground-based optical telescopes.

Unfortunately, NSF's Division of Astronomical Sciences was still reeling from paying for its share of ALMA, the Atacama Large Millimeter Array that was now operating in the Atacama Desert of northern Chile. At a cost of over $1.6 billion, ALMA has the dubious honor of being the most expensive ground-based astronomical telescope, eclipsing even ESO's relatively nearby VLT, which cost well over $400 million.

The Assessment noted that both the GMT and TMT projects had made "major progress" since 2010 and that both offered the ability to meet the science goals envisioned for the GSMT by Astro 2010, which included exoplanet detection and characterization. However (yes, the dreaded "however" was back), "programmatic hurdles remain," presumably referring to the Hawaiian Supreme Court, a formidable hurdle indeed, and "neither project has secured the funding need to complete construction at its full intended scope." Furthermore, NSF's budget constraints had prevented NSF from following the Astro 2010 recommendation to choose just one partner and to take that one to the dance. This was all summarized in a "Finding," a concise summary of the situation. Usually the Findings are followed by a relevant Recommendation or two. In this case, there was no such Recommendation, and the Midterm Assessment moved on to the next order of business. So what would NSF do about the GSMT commandment? Wait for Astro 2020?

Exoplanets in the Era of Extremely Large Telescopes (E3LT): The fourth annual GMT community science conference was held September 25–28, 2016, at the Asilomar conference center in Pacific Grove, California. I was limited in my usual ability to kibitz with other conference attendees in the back of the lecture hall while the meeting was underway because I had agreed to give the conference summary talk on the last day, and so I needed to listen closely to all the speakers and take good notes of what gems they revealed that might be worthy of showcasing a second time during my summary talk. I did pick up some insider information during the coffee breaks, however.

I asked a GMT insider about the status of the funding gap for GMT: was there any hope for selling off the remaining GMT time shares of the roughly $1 billion project? The answer was yes, there was hope. The back-of-the-envelope, and back-of-the-room, calculation, was that besides the existing commitments by the GMT partners for a total of $550 million, some of those partners were planning on increasing their time shares by raising another $200 million. In addition, several new partners were being sought among well-endowed, private American universities who had not yet joined either project, for an expected addition of another $200 million. Given the Hawaiian hiatus for TMT's construction schedule, it was looking good that the NSF would have to choose to pick the GMT as its dance partner to fulfill the Astro 2010 commandment to join a GSMT project. The GMT insiders were estimating that the NSF should be good for a 10% share, or another $100 million. That summed up to a grand total of about $1.05 billion, which just might be enough if further cost growth could be curtailed.

The TMT folks had not been showing any further interest in combining forces with GMT on the Las Campanas site but were betting the farm on either eventual success on Mauna Kea or else the backup plan for the Canary Islands. They would have to decide which one to choose by September 2017, if they wanted to stay in the race.

The E3LT meeting provided the first good opportunity since the public announcement of the reality of Prox Cen b to learn how planet hunters were reacting to this closest possible exoplanet. One of the first-light instruments for the GMT would be the G-CLEF (GMT-CfA Large Earth Finder) precision Doppler spectrometer, to be built by astronomers at the Harvard-Smithsonian

Center for Astrophysics in Cambridge, Massachusetts. This spectrometer would use an echelle grating to achieve high spectral resolution at optical wavelengths, with the goal being to achieve a precision of about 10 cm per second in the host star velocity variations, roughly the same as the 9 cm per second that the Earth induces in the Sun over a period of a year. Because the star Prox Cen is much lower in mass than the Sun (by a factor of about 8), and because the exoplanet Prox Cen b is a bit more massive than the Earth (by a factor of at least 1.3 times), the Doppler wobble of Prox Cen caused by Prox Cen b, about 140 cm per second, is considerably larger than that of the Earth on the Sun. That wobble should be detectable by G-CLEF, but given the noise associated with the star Prox Cen, it would still require a concerted effort to confirm the Doppler signal found by the PRD team.

E3LT also offered the chance to find out what some of the experts on direct imaging thought were the prospects for imaging this new world. It became clear that the GMT was being designed in such a way to permit the ultimate in excellent imaging from the ground. Each 8.4-m-diameter primary mirror would be paired with an off-axis, 1-m secondary mirror, in much the same way as was proposed long ago for TPF-C, the 3.5-m by 8-m optical space telescope. In addition, the six large mirrors in the outer ring of the GMT primary mirror assembly would not be obscured by any pieces of the support structure for the seven secondary mirrors, giving them each a clear view of the sky, without suffering the diffracted light that complicates high-contrast imaging when there is an intervening strut. Such an obscuration by the secondary mirror supports is intrinsic to the WFIRST optical telescope, as its original design was intended for imaging objects on the fully illuminated surface of the Earth rather than faint exoplanets next to relatively bright stars. The WFIRST secondary mirror is centered on the optical axis of the primary, and as a result, the six struts supporting the on-axis secondary mirror inevitably diffract and scatter the target star's light into the regions where one hopes to see the exoplanets. The WFIRST team of coronagraph experts was doing its best to overcome this diffracted and stray light originating from the spy satellite's peculiarly obscured optics, and they had achieved wonders; but on the whole, life is simpler if one does not have to start with an obscured primary mirror. GMT had made that critical design choice at the beginning.

Because each of the six outer GMT mirrors was conjugated to a single secondary mirror, this also meant that the large gaps between the six outer mirrors would not cause a problem with diffraction of light off the edges of the gaps. This approach differs considerably from that of the TMT, where the primary is composed of a hexagonal tile pattern of 492, 1.4-m mirrors coupled to a single, 3.6-m secondary mirror. Each of the gaps between the 492 primary mirror segments can be expected to diffract and scatter light toward the secondary, making a coronagraph's task of eliminating the star's light so that the planets may be seen that much more difficult. In addition, the TMT secondary mirror would be mounted on the optical axis of the primary so that its three struts would contribute as well to the extraneous star light. The GMT design would avoid these complications altogether.

There remains the formidable problem for ground-based telescopes of the Earth's atmosphere, which bubbles and brews even at superb telescopic sites such as Las Campanas and Mauna Kea. The solution is called adaptive optics (AO) in which the secondary mirror, and sometimes other optical components in the systems, are composed of a hexagonal pattern of rapidly deformable mirror segments, which tip and tilt, and piston in and out, in order to try to flatten out the distortions induced in the incoming lightfronts by the atmosphere's oscillations. GMT would use natural AO imaging from the very beginning, at the end of GMT's commissioning period around 2025. Natural AO imaging relies on having either a fairly bright target star, or other nearby bright reference star, in order to control the adaptive secondary mirror segments so that the smearing effects of the atmosphere can be removed. Natural AO imaging is a strength of the GMT team, as the University of Arizona astronomy team building the primary mirrors has already built and operated a highly successful natural AO imaging system on one of the 6.5-m Magellan telescopes at LCO (MagAO), and the team was developing an advanced version of extreme MagAO, termed MagAO-X, that would be a clear forerunner to what would be needed for the GMT (plausibly to be termed GMagAO-X).

So everything seemed to be in place for the GMT to begin operations in 2025 and to be able to perform the direct-imaging observations that would be the first to reveal the atmospheric compositions of the closest Earth-like exoplanets, right? Wrong. The first light instruments for GMT did not include a coronagraph, and it was unclear if one would be manifested for the second generation of GMT instruments, whenever that opportunity might arise. Oops.

Several direct-imaging concepts had been developed early in the GMT project and competed for becoming GMT's first light instruments, the instruments that would be ready to go when the GMT first opened its seven eyes. Two teams were led by Arizona's Phil Hinz and Olivier Guyon, and Alycia Weinberger and I were enthusiastic members of both teams. However, neither team succeeded in becoming a first light instrument, though Hinz's TIGER (Thermal-infrared Imager for the GMT providing Extreme contrast and Resolution) concept was considered and discussed at 2011's GMT Exoplanet Workshop as one of seven candidate instruments before being dropped.

None of the five accepted GMT instrument candidates was capable of even attempting to image Earth-like exoplanets, though one concept, GMTIFS (GMT Integral Field Spectrograph), would be able to image newly formed, hot, Jupiter-mass protoplanets orbiting at Jupiter-like distances from their young stars in the closest star-forming regions. That capability promised to yield important clues about the gas giant planet formation process, but it would shed little light on the formation of rocky worlds. The basic capability of imaging gas giant planets on wide orbits had already largely been achieved by ground-based extreme adaptive optics instruments, such as GPI on the southern Gemini telescope and by SPHERE on an ESO VLT telescope. As far as Earth-like exoplanets were concerned, though, GMT would be born blind.

It thus appeared that GMT would not be able to compete in the race to take the first images and spectra of Prox Cen b, leaving the race to be won by the TMT or the E-ELT, or possibly a NASA space telescope, once they became operational following the GMT.

In fact, it might not even take that long for Prox Cen b to be studied in some detail. Shortly before the Asilomar meeting started, Christophe Lovis of the Geneva Observatory in Switzerland and a number of his European colleagues submitted a paper to the European journal *A&A* that laid out their case for performing the first spectroscopic analysis of Prox Cen b using two existing instruments on the ESO VLT telescopes. While the SPHERE high-contrast imager could only detect objects brighter than about 1,000 to 10,000 times fainter than the host star, Prox Cen b might be studied if SPHERE could be combined with a new VLT spectrograph, dubbed ESPRESSO (Echelle Spectrograph for Rocky Exoplanet and Stable Spectroscopic Observations: European astronomers evidently favor pun acronyms referring to beverages such as Belgian beer or caffeine-laden coffee). Lovis and his team proposed building an interface between SPHERE and ESPRESSO that would allow the combination to just barely study the atmosphere of Prox Cen b. Even then, an enormous amount of VLT telescope time would be needed, perhaps 60 nights, spread out over 3 years, in order to possibly detect molecular oxygen. Water vapor and even methane might be observable as well. If nothing else, developing this VLT combination would be a step along the way toward building an E-ELT instrument that would certainly be able to perform this task. The E-ELT folks were already planning on building the Planetary Camera and Spectrograph (PCS), which would be a visible to near-infrared, high-contrast imager with more than enough resolving power to pick out Prox Cen b. PCS was planned to have first light in 2030. Clearly Prox Cen b was on everybody's wish list.

At the Asilomar meeting, Olivier Guyon noted that the GMT could do the job on Prox Cen b, if only it had a proper coronagraph to go with its AO system. The key point was that GMT was easily large enough to be able to resolve Prox Cen's Earth-mass planet by virtue of the effective 24.5-m diameter of its primary mirrors, over 10 times that of the Hubble-like NRO optics to be used for WFIRST. That large size would enable the GMT to image Prox Cen b, which orbited at a distance from its star of about 35 milliarcsecs (mas). One milliarcsec is roughly the angle subtended by a dime viewed from a thousand miles away. WFIRST would not be able to detect Prox Cen b, but with 10 times sharper vision, the GMT could do the trick.

In addition, a putative GMT coronagraph would not have to be any more efficient at detecting faint objects than coronagraphs that had already been built and tested in laboratories in anticipation of their use on WFIRST. This work had shown that planets that were 100 million times fainter than their host stars could still be detected and their spectra studied, with further improvements in this contrast ability to be expected. The ultimate goal is to image Earth-like planets, and the Earth is about 10 billion times fainter than the Sun at visible wavelengths. However, Prox Cen b is about 10 times fainter than the Earth, while the star Prox Cen is about 10,000 times fainter than the Sun, so that Prox Cen b might only be about 10 million times fainter than Prox Cen. In that case, a GMT coronagraph using proven designs and capable of achieving contrasts of 100 million to 1 should have no trouble picking Prox Cen b out of the glare of Prox Cen.

So clearly there was an urgent need to build a coronagraph for GMT, right? Besides taking the first photo of Prox Cen b, such a device would be able to use spectroscopy to detect key biosignature molecules in Prox Cen b's atmosphere, namely, oxygen, water, and methane. I ended my Asilomar conference summary talk on this high note in the hopes that the GMT folks would hear my plea, and followed this up in early 2017 with a response to a request to Carnegie Institution staff members for white papers on future strategic directions. I called for Carnegie and the GMT project to design and construct a coronagraph that would enable the first detection of biosignatures to be made on an Earth-like planet, Prox Cen b, by the GMT. The white paper was added to the stack submitted by other Carnegie staff members, awaiting a decision by the Carnegie Board of Trustees about their vision for the future direction of the Institution. Continued GMT participation seemed to be a given, but what about a GMT coronagraph?

News Flash—Another SpaceX Falcon 9 Mission Failure: In September 2016, a SpaceX Falcon 9 rocket exploded while being fueled with liquid oxygen for a prelaunch static test fire. The rocket and its satellite payload were destroyed, forcing SpaceX to once again halt further launches until such time as the cause could be ascertained and remedied. That meant several more months of launch slip for all of the satellites queued up to be launched by SpaceX for the next year or so. NASA's TESS mission was scheduled to be dispatched by SpaceX around December 2017, already delayed from the originally planned launch window of August 2017 by the previous SpaceX mission failure in June 2015. This second failure made it likely that TESS would not get off the ground until several more months later, perhaps in March 2018. Given that TESS was planned in large part to discover transiting exoEarths for follow-up spectroscopic characterization by JWST, and given that the JWST launch date of October 2018 was holding firm, there was beginning to be a question about just when TESS would be able to deliver an enticing target list for JWST. Although JWST's prime mission lifetime of 5 years was long enough to accommodate any likely TESS launch delay, it was beginning to look like JWST's very first exoplanet targets would not include any TESS targets. The TRAPPIST-1 system might be all there was to observe, at least to start.

Robbing Peter to Pay Paul: The Midterm Assessment had recommended in September that NASA APD increase spending for gravitational wave technology development, at the expense of further exoplanet imaging technology development beyond the WFIRST CGI, and that APD seek a larger role in the ESA L3 gravitational wave mission. We learned from Paul Hertz at the APS meeting on October 3, 2016, that L3 was on a faster track than had been previously planned, with a launch penciled in now for 2033 or 2034. NASA had been planning for a 10% share of L3, but there was a possibility that this share might rise to as high as 20%, the maximum fraction that ESA would permit for a partner in this mission. That share meant that NASA would provide on the order of $300 million to $350 million for L3; and given that launch was about 18 years away, this implied an annual cost of about $19 million, roughly double the previously planned APD support. It would take a few years to ramp up the L3 NASA spending from the planned program, but the threat was there in the long run to the rest of the APD mission portfolio, and the Midterm Assessment had already identified its preferred victim: exoplanet technology.

Given the October date and the annual federal budget process, any planned reprogramming of APD funds to address the Midterm recommendations would not become apparent until the FY 2019 Presidential budget request was released in February 2018—the FY 2018 budget plan had already been submitted to the OMB for their consideration. It was not clear how Paul Hertz would be able to pay for this unfunded mandate from the Midterm Assessment, but it seemed certain to delay further NASA's efforts to find and characterize habitable worlds: even WFIRST itself was not off the table.

Zurbuchen Arrives on the Scene: On October 3, 2016, Thomas Zurbuchen became the new NASA Associate Administrator heading up the SMD, filling the spot vacated by John Grunsfeld. Zurbuchen's strong background in university space science meant that SMD would now be headed by a genuine academic scientist rather than by a former astronaut with a doctoral degree. Good, but what did Zurbuchen think about the search for life beyond Earth? There was no need to worry about that. Zurbuchen made it clear from the beginning that the search for extraterrestrial life was a top NASA priority and that "major breakthroughs" could be expected in the next two decades. As head of SMD, Zurbuchen would be backing both the PSD's robotic searches for evidence of past life on Mars and the APD's planned space telescopes hoping to discern biosignature molecules on nearby worlds like Prox Cen b. Zurbuchen's Swiss origins made the "William Tell Overture" sound in my ears: the horse race was still on.

Keep the Faith: An email sent out to the Division of Planetary Science (DPS) members of the AAS on October 10, 2016, noted that the TMT folks would be holding a noontime workshop at the DPS meeting that fall. Oddly enough for such a workshop, which would normally be expected to focus on all the exciting science that such a major new facility would enable, and how one could become involved, the workshop was planned to provide an update on the status of the "search for an alternative construction site." Is that so? There was no mention of why an alternative site was being sought, as the assumption presumably was that everyone already knew. On a more positive note, the announcement noted that a free lunch would be provided, which is a sure-fire way to ensure that hungry scientists will attend in droves, just as cookies and doughnuts are routinely employed to lure graduate students and postdocs to attend departmental colloquia. Old habits die hard, especially when they are successful.

Hearings by the Hawaiian Board of Land and Natural Resources on the request by the TMT International Observatory to obtain a new construction permit for Mauna Kea were scheduled to start on October 18, 2016, on the Big Island. The expectation was that TMT would win another building permit, but the expectation was also that any new building permit would again be contested by native Hawaiian opponents and would again be referred to the state's Supreme Court for consideration. However, even if the Supreme Court should allow the project to proceed, the feeling was arising that Hawaii had simply become unsuitable for the TMT: local opposition could be expected to once again become intense when bulldozers began scraping away the sacred volcanic soil near the summit of Mauna Kea. The TMT board knew that they needed an alternative site and needed it fast.

The Gordon and Betty Moore Foundation had $180 million invested in the TMT so far, and this considerable clout had been directed toward keeping the TMT on Mauna Kea. Gordon Moore was the author of Moore's Law about the doubling of the number of transistors on integrated circuits every 2 years, and, not coincidentally, the co-founder of the Intel Corporation. The rumor was that the Moores had a home on the Big Island and wanted to be able to gaze up at the summit of Mauna Kea during the day and see tangible evidence of their philanthropy: a gigantic, 18-story-high dome, easily visible from the ground, or from airliners full of tourists passing by at 35,000 feet on their way to landing on Oahu. The TMT dome would be the largest structure on the entire Big Island of Hawaii.

By the end of October, the TMT board decided that enough was enough: the backup site for the TMT would be Roque de los Muchachos on La Palma in the Canary Islands. The TMT board decreed that construction must start at either Mauna Kea or the backup site by April 2018 at the latest in order to keep the project on track and competitive with the GMT and E-ELT. The ground-based race to image exoEarths was still on, with three strong contenders prancing around the paddock, saddled and waiting for their jockeys to mount.

The uncertainty attending the question of where to place the TMT had to be having a ripple effect on the other major issue facing the TMT: where to find the rest of the money needed to build such a monstrous telescope. A similar dilemma faced the GMT effort, though in that case the location had long ago been decided to be Cerro Las Campanas.

The E-ELT effort had the advantage of apparently full funding by the partner nations and a secure site on Cerro Armazones. Potential TMT donors and partners had be to leery of joining a hugely expensive, bricks-and-mortar science venture, where the land lease had not yet been settled.

14

November 8, 2016, A Date Which Will

God not only plays dice. He also sometimes throws the dice where they cannot be seen.

—Stephen W. Hawking, 1975, *Nature*, 257, 362

The United States awoke to the news that Donald J. Trump had won the Presidential election the previous day, winning more Electoral College votes than Hillary Clinton but losing the popular vote. Clinton had been leading Trump in the polls throughout the election campaign, and though the polls had become close toward election day, it was an enormous surprise to most Americans when it became clear that Trump would become the next U.S. President. But what did that mean for the search for habitable worlds? Science in general had not been mentioned much, if at all, by either campaign, and Trump had only offered an offhand remark about the need to fix the roads first when asked about NASA's future during a campaign visit to Florida's Space Coast.

Charlie Bolden immediately informed the NASA community by email that he expected the Trump Administration would "continue the visionary course on which President Barack Obama has set us." While Bolden's message was full of pride about what NASA had accomplished and was planning on doing, it was mostly whistling in the graveyard to me. We had very little idea what would happen with a Trump Administration. A Clinton Administration could be predicted to support NASA's ongoing work, but Trump's plans could not be so easily foreseen. A *Washington Post* reporter called me that morning to ask what I thought would be the likely effect of the Trump Administration on NASA and on space science in general. We had a short conversation, as I told her immediately that I did not have a clue. We would all have to wait and see, but not for long, as it turned out.

Predictions by Trump campaign advisors about the future of NASA began to appear later in November. They stated that the Trump campaign had a goal of shifting NASA funding toward "deep-space exploration." Combined with the antipathy of climate change deniers to NASA's Earth Science Division, with numerous satellites monitoring the Earth's atmosphere, and with the 50% build-up of NASA funds for the ESD during the Obama Administration, it was becoming pretty clear which of the four Divisions within the SMD was likely to feel the pain. It was even conceivable that the ESD might be dissolved altogether and its remaining satellites transferred to the National Oceanic and Atmospheric Administration (NOAA). Any funds thereby saved in the NASA budget might then be used for other endeavors, such as the Europa Clipper mission, or even for building new space telescopes capable of studying Earth-like exoplanets. NASA's Planetary Science and Astrophysics Divisions seemed likely to be spared any serious budget cuts. But it was hard to cheer these hints of what the future might hold for searches for the atmospheric components of an Earth 2.0 when cuts to the ESD budget would represent threats to the stability of Earth 1.0's climate. We do not really want to see what it would be like to try to live on a hotter, Earth 1.1 version of Earth 1.0. It seemed that NASA would soon face the results of decisions forced on it by Congressional or Administration leaders.

Representative Chris Van Hollen was elected to fill Barbara Mikulski's Maryland Senate seat. Van Hollen also replaced Mikulski on the CJS Appropriations Subcommittee. However, given that Senator Van Hollen was a junior member of the Committee, not the ranking member, his power might be diluted compared to that of the mighty Senator Mikulski. We would find out soon enough.

Crowdsourcing Eta_Earth: The crowd was getting restless and wanted to know just exactly what was Kepler's best estimate of the frequency of habitable Earths. One of the first activities I had when I was asked to chair NASA's ExoPAG in March 2015 was to consider a request from Rus Belikov, an astrophysicist at NASA Ames, to support the establishment of a new SAG. At that time, several astrophysicists had independently considered what the Kepler data released at various times was going to say about the value of Eta_Earth. These published estimates varied quite a bit, some by factors of 10 or more, and Belikov wanted to create a new SAG that would try to introduce some order into this process of estimating and extrapolating what the magic number should be based on what had been released by the Kepler project to date. The new SAG would consist of all the interested parties that Belikov could draw into the exercise, ranging from those who had already published their best guesses to those who were still working on the data as new releases were made. Needless to say this exercise seemed like a no-brainer to me: yes, it would be great to have all the experts on estimating Eta_Earth get together and try to converge on a single number, with error bars, if necessary. I had presented Belikov's proposed SAG charter to the NASA Astrophysics Subcommittee in July 2015, and the APS had immediately accepted the creation of this new SAG, coincidentally the 13th SAG.

By December 2016, Belikov's SAG 13 had been holding monthly telecons for over a year, seeking agreement on seemingly mundane, yet actually critical, parameters such as what radius range should be considered to be an Earth-size planet and what range of orbital periods for a solar-type host star should be considered to imply habitability. Once those definitions had been agreed on, and a dozen different approaches to estimating Eta_Earth had been recalculated to fit within the agreed-on ranges, it was possible to derive a best guess by simply giving each estimate equal weight and averaging them all together. By this means, Belikov came up with an estimate of Eta_Earth equal to about 60%. By astronomical standards, where differences by factors of two are often interpreted to imply agreement, SAG 13's estimate was essentially the same as Wes Traub's earlier estimate of unity. The implication was clear: *Earth-like planets are everywhere*. In order to clarify the source of this new estimate of Eta_Earth, I suggested that Belikov refer to his estimate as Eta_Earth_SAG13, which he agreed to do. Someday there might be an Eta_Earth_Kepler, agreed on and published by the Kepler team, or so I hoped.

Money, Honey: The new year 2017 broke with an announcement by the ESO that it had signed an agreement on January 9 to launch a targeted effort to use one of the 8-m VLTs to perform a search for exoplanets in the Alpha Centauri binary system. Yuri Milner's Breakthrough Starshot initiative would be putting up the funds to modify an existing mid-infrared imaging camera, VISIR (VLT Imager and Spectrometer for mid-Infrared), to "greatly enhance its ability to search for potentially habitable planets around Alpha Centauri." This meant that VISIR would need to have a coronagraph designed, built, and installed, along with an AO system to smooth out the distortions caused by the Earth's atmosphere. The deal also paid for a large block of VLT observing time, with the plan being to conduct a thorough search of the Alpha Centauri system in 2019.

Because habitable exoplanets give off significant mid-infrared radiation, being warm bodies themselves, they are not quite as faint at mid-infrared wavelengths compared to their stars as they are at visible wavelengths, where their light is simply reflected star light. As a result, the task of imaging a faint exoplanet next to a bright host star can be a thousand times easier to achieve using mid-infrared light instead of visible light. Factors of a thousand count for a lot in this challenging endeavor, and VISIR was decided to be the way forward for the VLT team. The major drawback to using these longer wavelengths of light is that the telescope must be twice as large in diameter in order to see as sharp a detail when using light with a wavelength twice as large; but given the 8-m size of the VLT primaries, Milner and the ESO folks were willing to give it a try.

The whole point of the Breakthrough Starshot initiative was to find a way to send mini-satellites to the Alpha Centauri system at speeds of 20% the speed of light, enabling them to reach Alpha Centauri in a mere 22 years or so, snap a few photos, and send them back to Earth, about 4.4 light-years away, at the speed of light, for a total elapsed time interval of about 26.4 years, gave or take a month or two. Time's a-wastin', as Snuffy Smith would say: let's find out where those mini-satellites need to go.

This new VISIR instrumental capability would come in handy for preparing for an even more ambitious mid-infrared imager and spectrograph that the ESO folks were planning on building for E-ELT, called METIS (Mid-Infrared E-ELT Imager and Spectrograph). While METIS would be far more capable than the enhanced VISIR on the VLT, obviously the race was on even within ESO to beat itself to the punch in getting the first images and spectra of a nearby habitable world: 2019, after all, was effectively "now," compared to the planned first light of the E-ELT of 2024. Time is indeed a-wastin'.

The discovery of Prox Cen b was also noted in the press release as a motivating factor, and it seemed clear that this fortuitous discovery, announced well after the initial Breakthrough Starshot initiative about possible Alpha Centauri exoplanets, must have become a concrete motivating factor for this VLT VISIR investment: the Milner team now knew for certain that there was at least one habitable Earth-size world in the Alpha Centauri triple-star system, if not in the binary itself, with its hot exoEarth around Alpha Centauri B. Prox Cen b would be the likely target for the Starshot mini-armada.

The List Goes On: Short of a visit by a bevy of Breakthrough Starshot mini-satellites that sends back unambiguous close-up photos of an exoplanet with oceans of blue-green algae, forests, and meadows full of chlorophyllic planet life, or a Great Wall of ExoChina, astronomers must rely on the search for biosignature molecules in the spectrum of an exoplanet's atmosphere to make the case that the planet is inhabited, not just habitable.

The four classical biosignature molecules are carbon dioxide, water, oxygen, and methane. Carbon dioxide represents about 96% of the atmospheres of Venus and Mars, and is present at the 0.04% level in Earth's atmosphere. Water can be as abundant as 5% on a humid day on Earth. Water is present at trace levels (about 20 ppm) on Venus and is locked up as frost and ice on the cold Martian surface. Oxygen is a major fraction, about one-fifth, of Earth's atmosphere, but it is only present at the 0.15% level on Mars. Methane is the rarest molecule of these four biosignatures, occurring at the 0.0002% level on Earth and in traces of parts per billion on Mars. Methane, though, is the gold standard for biosignatures, at least when it occurs along with oxygen. When both of these molecules are present, there must be a sustained supply of methane to replace that lost steadily by chemical reactions with oxygen.

The fact that trace amounts of methane have been detected on Mars, and the fact that Mars does have oxygen at a not insignificant level, is one of the best arguments, if not the best, that Mars might have some sort of subsurface life capable of producing methane, just as methanogenic bacteria do on Earth. Amazingly, it is estimated that roughly one-third of Earth's steady supply of methane is derived from the exhaust gases produced by the intestinal systems of ruminants such as cows, goats, and sheep. While the Mars rovers would be more than pleased to snap a photo of a cow's fossil skeleton half-buried in the Martian soil, the fact that methane is detected on present-day Mars, coupled with the absence of any obvious extant life forms above the Martian surface, has led many to search for abiotic sources of methane, that is, nonbiological chemical reactions among the minerals and rocks near the surface. Mars 2020, the next NASA mission to Mars (scheduled for launch in, you guessed it, 2020) has been planned to carry drills that will allow the rover to probe beneath the rock surface layers in search of whatever is producing the methane, be it biotic or abiotic processes. In addition, some of the more promising materials recovered from the drilling will be stored in tubes and cached, for later retrieval and transport to Earth laboratories by a future NASA mission.

Searching for and finding evidence for the existence of past or present life beyond Earth is extremely difficult, even when the subject planet is as readily accessible and well explored as Mars. Suffice it to say that the problem becomes exponentially harder when the planet is multiple light-*years* away, instead of multiple light-*minutes.* (At the closest approach to Earth, it takes about 4 minutes for a radio wave to travel from Mars to Earth, and vice versa. So don't hang up if you don't hear a reply immediately. Please be patient. Thanks.) This means that biosignature molecules are the only real-time means for discovering evidence that might support an argument for extraterrestrial life on an exoplanet, as contentious as that might be. The light carrying those faint hints of biosignature molecules may have left the exoplanet atmosphere many decades ago, but that light is still streaming steadily past the Earth, waiting for us as a species to become intelligent enough, curious enough, and proficient enough to build telescopes capable of its interception and decipherment. The universe is patient. It has waited about 13.8 billion years so far for us to do this, so waiting a few more years is no problem, no problem at all.

To help matters along, in late 2016, Sara Seager of MIT (see Figure 14.1) and her colleagues published an article in the journal *Astrobiology* that produced a master list of gas molecules that should be considered as potential biosignatures beyond the classic four molecules. Taking the philosophy that all volatile gas molecules should be considered as possible contenders, they produced a list of about 14,000 molecules, of which about 2,500 contained the elements usually associated with life and organic chemistry: namely, carbon, nitrogen, oxygen, phosphorus, sulfur, and hydrogen. Of these 2,500 molecules, Seager and her team showed that about one-quarter were known to be produced by terrestrial life. Even this long list was restricted to molecules that had no more than six atoms that were not hydrogen atoms. That meant that there were now at least 600 molecules that might be considered as potential biosignature molecules. Gadzooks. Would any of those molecules be detectable in an exoplanet's atmosphere, at least by any telescope we could afford to build?

Preparing for the Big One: Studies of the four large-mission concepts being developed for consideration as APD's flagship mission for the late 2020s were well underway in February 2017, and JPL's optical engineers were working on preparing for the development activities that would be necessary to raise the technology readiness level (TRL) of any new large-mission concepts from the bottom level of 1 (meaning a sketch on a napkin during a cocktail hour) toward the top level of 9 (meaning the concept had been thoroughly developed, tested, and flown successfully). Technologies with TRL values in the range of 5 to 6 were generally considered

Figure 14.1 Sara Seager, a leader in the field of exoplanet atmospheres and spectroscopy (Courtesy of DTM, Carnegie Institution for Science).

to be at the minimal end of the safe range for serious proposals for NASA missions, and considerable work would be needed to get any new ideas for the large missions to these TRL values.

The High Contrast Imaging Testbed (HCIT) at JPL would continue to be an important means for demonstrating the effectiveness of various optical designs for coronagraphs for achieving the ability to photograph faint exoplanets close to their host stars. The HCIT had already been instrumental (pun intended) in showing that several of the coronagraph designs under development for the WFIRST space telescope CGI were at a high enough TRL level to be further improved once WFIRST entered the next phase of mission development. In the meantime, the HCIT folks were proactively seeking to aid the HabEx and LUVOIR large-mission STDTs in their deliberations by planning to upgrade the HCIT to become the modestly named "HCIT Decadal Survey Testbed," with the goal being to demonstrate the ability to reach contrasts of 10 billion to one so that an exoEarth that was 10 billion times fainter than its host star in visible, reflected light could still be discerned in the blinding glare of the star. This upgraded testbed would not only test coronagraphs applicable to the presumed HabEx base case of an off-axis, monolithic primary mirror, the theoretically ideal configuration for high-contrast imaging, but also coronagraphs that would build on the work performed for the effectively segmented, on-axis WFIRST primary, in order to consider primaries of the sort expected to be proposed by the LUVOIR STDT, namely, progressively larger versions of the JWST primary mirror, with its 18 hexagonal segments. The upgraded HCIT was planned to be ready to go once the HabEx and LUVOIR STDTs released their final reports; and then the race would be on to see if any low TRL concepts could be raised high enough to be taken seriously by Astro 2020.

NASA Watch Figures It Out Ahead of Time: I spent part of the morning of February 22, 2017, talking with a journalist who was looking for more information about what would be presented at a NASA press conference scheduled for 1:00 PM. The press announcement had only stated that the topic would be something new about exoplanets, with the details left out because the topic was presumably the subject of a paper about to be published by a journal that insisted that its embargo on hot new results be honored by all. Normally I would not necessarily have any idea about what was about to be presented, unless the journalist had an advance copy of the paper and was willing to send me the embargoed copy for my comment; but in this case, the web site NASA Watch had figured out the story the day before. The NASA press announcement had noted that Michael Gillon would be one of the participants, and NASA Watch's Keith Cowing took it from there. Gillon had already achieved a major success by announcing on May 3, 2016, the discovery of three Earth-size planets in orbit around TRAPPIST-1. Cowing did some detective work and figured out that Gillon was about to announce the discovery of four more Earth-size planets in the TRAPPIST-1 system, for a total of seven such worlds. Several of them, perhaps two or three out of the seven, might very well orbit at distances where liquid water would be stable on their surfaces.

The TRAPPIST-1 planets were all discovered by transit photometry, that is, by watching the light from the low-mass, red dwarf star dim periodically as each of the now seven planets passed in front of the star. That meant that all seven planets were orbiting in a single, thin orbital plane, a plane whose orientation was fortuitously aligned with the direction toward Earth. Mother Nature could not have given us all a better Valentine's Day present, even if it was delayed a week by the journal *Nature*'s press embargo.

The earlier *Nature* article about the first three planets had been dramatic enough that Gillon and his colleagues were able to convince the Spitzer folks to spend an extraordinarily long 20 days staring at TRAPPIST-1 in order to refine the timing of the transits of the previously known three exoplanets and to search for however many more could be found in the same orbital plane. Ongoing monitoring by the now two TRAPPIST telescopes, Spitzer, the ESO VLT, and an astounding variety of other telescopes around the world meant that dozens of co-authors shared the credit for this remarkable discovery. The fact that the TRAPPIST-1 story did not leak out before Cowing did his investigative journalism meant that the TRAPPIST-1 team was considerably more leakproof than the Trump White House was proving to be, just one month into the new Administration.

Given the quality of the Spitzer transit timing observations, the second *Nature* article was able to make good estimates of the mass of six of the seven new exoplanets.

This was accomplished by a technique developed by Matt Holman for Kepler data called transit timing variations, or TTV. The inner six planets of the TRAPPIST-1 system were so closely packed together, all within about 0.03 times the Earth–Sun distance, that the gravitational forces between them led to small variations in exactly when each successive transit of a given exoplanet was detected by Spitzer: sometimes the transit would occur a bit earlier than on average and sometimes a bit later. The fact that the inner six were orbiting in a synchronized dance, with their orbital periods close to being in resonance with each other, meant that small gravitational tugs on each other would occur over and over again, amplifying the TTV signal. When analyzed, the TTV signals implied that six of the exoplanets have masses less than about 40% greater than that of the Earth. The transit depths had shown that these six have radii less than about 10% larger than Earth. Clearly, TRAPPIST-1 has at least six truly Earth-size planets.

An accompanying commentary in *Nature* noted that the discovery of the seven-planet TRAPPIST-1 system meant that "our Galaxy could be teeming with Earth-like planets." Although that conclusion had already been drawn by many from the Kepler Mission data, this new discovery could only lend further motivation to the race to search for habitable worlds. It was immediately clear that because the TRAPPIST-1 planets were all transiting planets, this system would undoubtedly be one of the first to be studied in detail by JWST, whose strength in exoplanet characterization was limited largely to transit spectroscopy. JWST had not been designed to perform imaging of nearby exoEarths, even though that had been one of the two major goals of the mission concept that evolved into JWST. (The original mission concept was published in the *HST and Beyond* report, released in May 1996, soon after the October 1995 announcement of 51 Peg b, as discussed in TCU. The editor of *HST and Beyond,* and chair of the committee, Carnegie's Alan Dressler, was awarded the American Astronautical Society's 2017 Carl Sagan award in recognition of the committee's work in laying the foundation for JWST.) TRAPPIST-1 was a transiting system perfect for JWST's spectroscopic strengths, with multiple Earth-like, potentially habitable exoplanets, ripe for thorough examination. NASA ExEP managers and JWST scientists were so thirsty for the TRAPPIST-1 planets that they could already taste the water that JWST might find there.

The TRAPPIST-1 system was a bit far afield, though, for the tastes of the Breakthrough Starshot crowd, being almost 10 times farther away than the Alpha Centauri system. Even at 20% of the speed of light, that meant a travel time of at least two centuries instead of two decades for the mini-spacecraft fleet. The two *Nature* articles had quoted the same distance for TRAPPIST-1 of 39 light-years, and the latter article did not cite the CAPSCam team's 2016 ApJ article giving an improved distance of 41 light-years. Precise distances are important for determining the parameters of the host star, which in turn are important for determining the properties of any transiting exoplanet; so a 5% increase in the distance to TRAPPIST-1 was not as negligible as it might seem. Alycia Weinberger and

I shrugged our shoulders when we saw the new *Nature* article, but this time I sent Gillon an email of congratulations that evening, along with a note that the DTM CAPSCam group had published an updated distance for the TRAPPIST-1 system that he might wish to consider. Gillon graciously replied his thanks shortly thereafter, in spite of the deluge of email and press attention he was justly receiving.

The NASA Exoplanet Exploration Program folks were ecstatic about the press attention generated by this discovery: the social media reaction by Twitter and Facebook users was the largest response to any NASA event for the last several years. One wondered if President Trump, who was becoming known as the nation's leading Twitter devotee, happened to see the TRAPPIST-1 media storm.

President Trump's First Address to a Joint Session of Congress: On February 28, 2017, I was frantically trying to find a place to park close to the Cannon House Office Building, where I was slated to speak to the House Earth and Space Science Caucus at 2:00 PM. There was no place to park, on or off the street: Capitol Hill was flooded with lobbyists, standing outside in long lines, waiting to clear the security checkpoints and enter the various House and Senate office buildings. President Trump would be addressing Congress and the nation that evening, and as a result the federal security was already considerably tighter than it normally is on Capitol Hill. The lobbyists were out in full force—the moment was now to make a case for their clients. I ended up parking a half-mile away, across from NASA HQ, where from long experience I knew there was a good parking garage, and hustled back to Capitol Hill just in time.

The briefing went smoothly, though given the considerably more pressing events occurring on Capitol Hill that day, it was attended chiefly by junior staffers and interns looking for a free diet soda and a handful of popcorn. On the way back to retrieve my car, I stopped by NASA HQ, where planetary scientists were holding an open workshop to consider plans for what the NASA PSD might accomplish in the next several decades. The Vision 2050 Workshop was being held in the ground-floor James Webb auditorium, outside the building's security perimeter, and so it was often used for public events and press conferences. The Webb auditorium was packed to the gills with planetary scientists and their grand dreams for exploring the remaining nooks and crannies of our Solar System. But the real question that day was what in the world President Trump would say with regard to NASA's future during his speech that evening. Would NASA even be mentioned at all? The rumors were beginning to fly that Trump would indeed mention space that night, but only in the sense of calling for more resources for human space flight. In an era in which federal agency budgets were declining, or flat at best, that could only mean that something else in NASA's budget would have to be deleted or downsized to pay for increased spending in the human space flight side of the house. No one was about to volunteer to cut their budget to help out.

President Trump went for the bleachers in his Congressional address, saying that "American footprints on distant worlds are not too big a dream." Uh, Mr. President, exactly which "distant world" did you have in mind? That would have been a wonderful follow-up question to ask, but of course there was no opportunity to pose such a question that evening. The Moon? An asteroid? Mars? Europa? Prox Cen b? Trump was a Big Picture guy and evidently would leave the details to others to figure out. But who? Charlie Bolden had tendered his resignation as NASA Administrator with the change in Administrations on January 20, as political appointees are expected to do, but the new Administration had yet to propose a replacement for Bolden. Acting Administrator Robert Lightfoot did his best to rally the NASA troops, sending out an email to all NASA employees on March 7, 2017, noting that in spite of the President's FY 2018 budget calling for a $54 billion increase in defense spending, to be offset by a $54 billion decrease in discretionary spending, "we remain confident in the Administration support for NASA." Lightfoot noted in passing that the Congress, of course, would have the final say on the FY 2018 budget. Nudge, nudge, wink, wink.

The same day, March 7, the House approved the Senate's version of a joint bill, the NASA Transition Authorization Act of 2017. This bill was intended to reaffirm bipartisan, bicameral Congressional support for NASA. Representative Lamar Smith stated that the bill would "take another step in making America great again," echoing the Trump campaign's political catch-phrase. Many of us thought that America had never stopped being great, but if someone wanted to make America even greater, who were we to argue with that? The House press release pointed out that the Act called for continued funding for JWST, WFIRST, and the Mars 2020 rover and called for more study of a Europa mission, in addition to how to develop the technology for a human mission to Mars. One could not have asked for a better path forward for "making America great again." Okay, fine, show us the money. The FY 2018 budget was now in Congress's hands.

The Transition Authorization Act also hinted that Congress was quite aware of the upcoming Astro 2020 Decadal Survey. Although not directly linked to Astro 2020, the Act called for NASA to work with the NAS to commission a "science strategy for the study and exploration of extrasolar planets," specifically as this strategy related to the capabilities of the upcoming TESS, JWST, and WFIRST space telescopes. The Congressional language then referred to "or any other telescopes, spacecraft, or instruments," presumably meaning, let's see, TMT, GMT, E-ELT, CHEOPS, PLATO, Gaia, and a host of ongoing, ground-based Doppler spectroscopy and transit photometry operations, such as the TRAPPIST effort, not to mention the heroic ongoing work performed by Spitzer, Hubble, and Kepler/K2. Did I miss anything? The Report would have to sift through an huge array of present and future exoplanet observational capabilities.

The clock was ticking. The Report was due to be submitted to NASA and Congress within 18 months after the Act became law. The date on the enacted legislation was given as March 21, 2017, so that meant submission was required by September 21, 2018, give or take a day. That timing meant that the Report would be ready for submission to Astro 2020, which was slated to begin work in early 2019.

Who would be tapped by the NAS to participate in concocting this possibly key planning document? What would they decide? Would the Astro 2020 committee even pay much attention to the Report's conclusions, given Astro 2020 would have to consider the priorities for all of astronomy and astrophysics, not just for exoplanets, as sexy as they might seem?

15

And the Winners Are

Physics is much too hard for physicists.

—David Hilbert, 1862–1943

The front page of the *Washington Post* on March 16, 2017, listed the winners and losers among the federal agencies participating, whether they wanted to or not, in the first "Make America Great Again" Presidential budget. The FY 2018 White House budget proposal requested over $19 billion for NASA, a 0.8% decrease in the total budget, compared to the FY 2017 total, still under a continuing resolution. Most other non-defense agencies fared considerably worse in comparison, with some facing cuts of 10% or more. It used to be said that flat was the new up, in terms of dismal federal budget prospects in the era of massive federal deficit spending; now it had to be said that a 0.8% decrease was the new up. That morning, Robert Lightfoot emailed his NASA employees to point out that the new budget top line meant that NASA would continue to be able to do its work, and that "overall science funding is stable," though there would have to be some science missions in development that would "not go forward." Evidently, that was the new euphemism for "dead." Perhaps not surprisingly, Lightfoot stated specifically that the mission to capture and return a sample from an asteroid was now dead, a mission that had been championed by President Obama. President Trump was now large and in charge, and most of the Obama team was long gone. But what else was hidden in the President's FY 2018 budget? Lightfoot's email gave few clues.

Later that morning, I was able to download the OMB's FY 2018 budget document, entitled "America First: A Budget Blueprint to Make America Great Again." Yes, indeed, President Trump was now in charge. The NASA section did not have much detail, but it did specify that the Mars 2020 rover and the Europa flyby mission were still very much alive. However, because of the need to maintain program balance (i.e., not hand the checkbook to the PSD), the Europa lander mission was not being funded. More dramatically, four upcoming Earth climate satellite missions were being "terminated." So much for the pleasantry of missions "not going forward": the OMB was in no mood for euphemisms. At least they did not use the covert intelligence agency phrase "terminate with extreme prejudice."

That was about all that the OMB was willing to show of the hand it was dealing to NASA. There was no mention of the APD, or of its two flagship missions, JWST and WFIRST. Was the OMB betting on PSD's hopes of finding life on Mars or Europa over APD's search for life beyond the Solar System?

Who Wants a Probe?: In preparation for Astro 2020, the APD held a competition for supporting teams that wanted to develop a probe-class mission concept, something spectacular that could be done for less than a billion dollars. After a peer review considered 27 proposals, 10 were chosen for funding 18-month-long studies. The March 20, 2017, announcement stated that 2 of the 10 would be exoplanet studies: one for a space telescope designed to perform high-precision Doppler spectroscopy above the Earth's atmosphere, and one for a star shade that would rendezvous with WFIRST. The latter study, led by Sara Seager, would be an update of the Exo-S star shade rendezvous mission study performed back in 2015. The 18-month time span ensured that the final probe-class mission reports would be ready for submission to Astro 2020, once it opened for business in early 2019.

Next Stop, Prox Cen b: The Breakthrough Starshot crowd held a scientific conference at Stanford University on April 20–21, 2017, that focused on Earth-like planets in the Alpha Centauri and the TRAPPIST-1 systems. Appropriately enough, Guillem Anglada-Escude and Michael Gillon would be the keynote speakers; and, as expected, it was announced that the primary target for the Starshot armada would be, you guessed it, Prox Cen b.

With an excellent sense of timing, on April 20, 2017, *Time* magazine had released its list of the 100 Most Influential People in the world for 2017, a list that included, besides Anglada-Escude and Gillon, Natalie Batalha, who had become effectively the face and voice of the Kepler Mission after Bill Borucki retired from NASA in July 2015. The three astronomers were included in *Time*'s subgroup of the 25 Most Influential Pioneers, of which the only names I recognized besides Natalie's, Guillem's, and Michael's were those of Samantha Bee, Ivanka Trump, Jared Kushner, and Jordan Peele, a mixed group, to put it politely. It was odd to think of equating the achievements of pioneering exoplanet astronomers, whose discoveries would be heralded in textbooks, with those of pioneering comedians and pioneering political novices, but there it was. It could not be argued that they were not an influential bunch. *Time* got that right.

A few days after the Stanford Breakthrough conference, Michael Gillon spoke at NASA HQ about the discovery of the TRAPPIST-1 system and his plans for the next steps. On April 24, 2017, the NASA Astrophysics Advisory Committee (APAC) was holding its first meeting since the APS was converted to a full FACA entity. This major change primarily meant that we could now submit our remarks directly to the person who most needed it, Paul Hertz, APD Director, rather than follow the cumbersome procedure for giving advice that characterized the APS—namely, passing APS remarks up to the NAC Science Committee, which could then pass the remarks up to the NASA Advisory Council, which could then approve the remarks and pass this advice to the Administrator, who could then pass it back down the chain of command to the SMD Associate Administrator, who would let Paul Hertz know about something that he had already learned about several months earlier at the APS meeting. Needless to say the previous reporting procedure was not a model of responsiveness and clarity of message and reminded one of the party game where someone whispers a message in another's ear, who does the same to the next person, and then after the message has been sent and received a dozen or so times, the resulting garbled version can be compared with the original. The old APS reporting system was a marvelous means to minimize the flow of gratuitous advice within the corridors of NASA HQ; the new APAC designation meant that Paul Hertz could get a direct answer to any questions he chose to pose to our FACA committee.

Whereas most of us on the APAC knew all about the TRAPPIST-1 discoveries, Gillon also informed us about something called SPECULOOS, yet another forced acronym, standing in this case for "Search for habitable Planets EClipsing ULtra-cOOl Stars." While SPECULOOS had the advantage of not standing for yet another beverage, my suspicion was that the acronym must stand for something European. It was only by Googling the acronym that I learned that "speculoos" is a traditional Belgian spiced cookie, and that in fact it has the same recipe as that for the Dutch "speculaas," or windmill cookies, which my Frisian grandparents enjoyed with their daily afternoon tea.

SPECULOOS was planned to be a new transit survey that would find more potentially habitable exoplanets that could be examined by JWST for signs of life. SPECULOOS would be a group of four robotic 1-m telescopes to be installed at ESO's Paranal Observatory. Gillon planned to use SPECULOOS to search methodically through about 1,000 so-called ultra-cool dwarf stars, stars that are low enough in mass to have surfaces cooler than about 2,700 degrees Kelvin; for comparison, the Sun's effective temperature is about 5,800 degrees Kelvin. TRAPPIST-1 was one of the first ultra-cool dwarfs to be searched for transiting exoplanets, and evidently Gillon was not content to rest on his TRAPPIST-1 laurels. He estimated that for an Earth-twin exoplanet orbiting in the HZ of a star at a distance of 10 parsecs (32.6 light-years), JWST would be able to detect the presence of the biosignatures water, ozone, and carbon dioxide in its atmosphere, with a high signal-to-noise ratio, in a mere 200 hours of JWST mid-infrared observations, provided that the star was an ultra-cool dwarf. JWST might also be able to make these detections for a sun-like star, but it would be harder to do because of the much brighter primary star light compared to that for an ultra-cool dwarf, and so would require an even larger block of precious JWST observing time. Gillon pointed out the need to keep the SST functioning, as Spitzer had played the key role in discovering the additional planets, and their masses, in the TRAPPIST-1 system.

It sounded like Gillon's SPECULOOS project was planning on poaching on the turf of the TESS mission, which was also intended to discover potentially habitable, transiting planets for follow-up by JWST. In fact, it was not. By virtue of its very name, SPECULOOS was focusing on the ultra-cool, faint dwarfs lower in mass than the red dwarfs to be monitored by TESS. Where TESS dropped off, SPECULOOS would pick up, as the stellar mass decreased, making for a complementary transit search.

With four, 1-m telescopes, SPECULOOS would be able to search fainter M dwarf target stars than the similarly motivated "MEarth" survey already underway in the Northern Hemisphere at Mount Hopkins, Arizona, and at the Cerro Tololo Inter-American Observatory (CTIO) in Chile, which consisted of eight, 40-cm (16 inch) telescopes at each site. The MEarth team had just announced their third exoplanet discovery, around a red dwarf star called LHS 1140. Their article in the April 20, 2017, issue of *Nature* took direct aim at the Prox Cen b discovery, noting that it did not appear to be a transiting planet, and at the TRAPPIST-1 system, claiming that those seven planets had "poorly constrained" masses. LHS 1140 b, on the other hand, was trumpeted as a rocky planet with a mass 6.6 times that of the Earth, just 12 parsecs away, orbiting in the star's HZ. As a transiting super-Earth, LHS 1140 b was certain to be high on the observing list for JWST; but its 12-parsec distance, similar to that of the TRAPPIST-1 system, meant that the Breakthrough Starshot team would not be changing their mind about heading for Prox Cen b. Queried by an Associated Press reporter about the breathless LHS 1140 b press release, which claimed that LHS 1140 b had "catapulted itself to the top" of the list of "the best planets outside our Solar System to look for signs of life," I replied that I put LHS 1140 b in the number-three spot, given the closeness of Prox Cen b and the presence of seven transiting planets in the TRAPPIST-1 system. In the language of the horse track, I was betting on Prox Cen b to win, the TRAPPIST-1 system to place, and LHS 1140 b to show.

Clearly Gillon wanted SPECULOOS to jump squarely into the race to find the best exoplanet candidates for JWST to characterize: not only was he planning the Southern Hemisphere search at Paranal but also a Northern Hemisphere search, just as the MEarth survey was doing. Competition abounds in the search for exoplanets: sometimes complementary, sometimes head to head. Who would get the blue ribbon for finding the grand prize, the first habitable world with clear evidence of biomarkers?

At the APAC meeting (April 24, 2017), we learned about launch dates for the upcoming exoplanet missions. TESS had indeed been rescheduled to fly in March 2018 on a Falcon 9, just 6 months before JWST's Congressionally mandated October 2018 launch window opened. WFIRST was planned for launch in the mid-2020s and was scheduled to move into Phase B of mission development in October 2017. The WFIRST team was instructed to continue to plan for maintaining the capability of operating along with a star shade, should one be highly ranked by

Astro 2020, with a final decision about star shade compatibility to be made by the fall of 2017. The European Euclid mission, a competitor to WFIRST in the area of dark energy studies, was still not planning on having a microlensing exoplanet survey as a prime mission objective, unlike WFIRST. Euclid was now shooting for a 2020 launch, but the absence of any major exoplanet goals would make its relatively early launch date moot in the race for habitable worlds: Euclid had effectively been scratched from the race long ago.

Not So Fast: The plan to move the WFIRST mission development into Phase B in October 2017 hit a snag on April 27, 2017, when it was announced that the project would be subjected to the WIETR: the WFIRST Independent External Technical/Management/Cost Review. Yikes. After the JWST debacle, NASA had wisely become rather sensitive with respect to the costs and schedule risks associated with flagship missions. WFIRST was the next flagship in the queue after JWST and would be the first beneficiary of the lessons learned from the development of JWST. WFIRST's estimated cost had risen steadily for the last several years to a current estimate of about $3.2 billion. It made eminent sense to blow the whistle on the WFIRST game, call a timeout, and have an independent team of experienced mission developers and scientists come in and take a close look at what the WFIRST team had accomplished during Phase A and where they planned to go in the second quarter.

The WIETR had been undertaken by the NASA APD in response to a recommendation by the Midterm Assessment of Astro 2010 to perform such a task. The terms of reference for the review focused on four items, three of which were derived from the acronym itself—namely, the technical requirements, the management processes, and the cost and schedule risks. The fourth, however, was much more specific: the coronagraph that was the shining jewel in the WFIRST crown of exoplanet discovery. That term of reference asked a pointed question: "Are the benefits of the coronagraph to NASA objectives commensurate with the cost and cost risk of development?" This was seemingly a reasonable question to ask, though it was not asked specifically of WFIRST's other science goals, as those had already been blessed in perpetuity by Astro 2010. It was just the CGI that was in mortal danger, pending the outcome of the WIETR review. The d-word had appeared in the terms of reference: it was now formally possible that the CGI might be descoped from WFIRST if the WIETR gave it a thumbs down.

The schedule for the review was an ambitious one: just 8 weeks from start to finish. If completed successfully and on time, the WIETR interlude would be a relatively painless intermission between Phase A and Phase B. The WIETR press release stated that NASA intended to appoint the WIETR panel members by the end of May, so that the final report would be delivered by the end of July 2017, if things remained on track and on schedule. A timely review would be important for keeping the total cost of WFIRST under control: the timecards of the scientists and engineers working on designing and developing WFIRST were still being filled out each week, whether or not Phase B had been reached.

Congress Just Says No: On April 30, 2017, the House and Senate finally agreed on the top budget numbers for the current fiscal year, which had begun 7 months earlier. Their budget proposal for FY 2017 would reverse most of the cuts to science funding that had been included in the President's budget for FY 2018. NASA would get $19.7 billion, an overall increase of $400 million over the FY 2016 budget total and $600 million more than the Trump FY 2018 request. In particular, planetary science made out like a bandit, getting $300 million of the overall $400 million increase, enough to plan on both the Europa orbiter and lander, as well as Mars 2020. The latter mission planning might even consider sending a helicopter along for the ride to Mars. Say what? Hopefully the use of the word "helicopter," which conjures images of Marine One, the large Sikorsky SH-3 helicopters used to ferry the President from the White House to Joint Base Andrews, was not really what was intended for Mars: perhaps an unmanned drone would be closer to describing what NASA might be able to afford to piggyback to Mars along with the next rover mission. There was no word in the budget deal announcement about how JWST, WFIRST, and the APD in general had fared; one could only hope that no news was at least not bad news.

Late on May 8, 2017, Robert Lightfoot emailed NASA's employees to let them know that President Trump had signed a bill a few days earlier giving NASA a FY 2017 top number of $19.6 billion, a mere $100 million less than the number cited by the Congressional press release. Perhaps $100 million was considered in the noise, merely a rounding error, but we would learn more when the details were released. Lightfoot did mention that JWST was still on track for launch in 2018, and there was a "strong overall science portfolio," so that boded well. We needed to get JWST off the Earth and on its way to L2, where it belonged, so that NASA could afford to build WFIRST next.

Was George Gatewood Right After All?: Scrolling through the articles published in the May 2017 issue of the *Astronomical Journal* (AJ), I noticed a remarkable article published by my colleague down the hall, Paul Butler, which I hitherto had not known existed. Paul's AJ article presented a master tabulation of the results of the Doppler exoplanet surveys he and his Lick Observatory and Carnegie colleagues had been carrying out over the last 20 years using the HIRES instrument on the Keck I telescope on Mauna Kea. The tabulation covered 1,624 stars, of which 357 were considered candidates for exoplanet signals. Of the latter, 225 had already been published as being authentic; but another 60 were still in limbo, awaiting final confirmation by follow-up studies such as photometric monitoring of the star's activity cycle, a possible source of spurious Doppler variations. Another 54 still needed more Doppler observations to become candidates.

The line in the AJ article's abstract that caught my eye the most was the detection of a planetary candidate around Lalande 21185. Lalande 21185? Really? That was a shocker to me. As I described in detail in LFE, Lalande 21185 had been claimed to host a planetary system back in June 1996 by George Gatewood, the astronomer who had disproven in 1973 the claim made by Peter van de Kamp (in 1963) for an astrometric detection of a giant planet around Barnard's star (also detailed in LFE). By 1996, Gatewood's astrometric observations of Lalande 21185, coupled with the exoplanet fever resulting from the seminal announcement of 51 Peg b the previous fall, led him to announce that Lalande 21185 has not one, but *two* giant planets, orbiting with periods of 6 and 32 years.

In spite of the large number associated with the name, Lalande 21185 was a star of prime importance to astrometers, as it is the fourth-closest star to the Sun, only 8.2 light-years away. Lalande 21185 was merely the 21,185th star on the list of 50,000 stars compiled by the French astronomer Joseph-Jerome de Lalande in 1801, but it was close by, and that made it a prime candidate for astrometric planet searches such as that of Gatewood at the Allegheny Observatory in Pittsburgh. If one counted the Alpha Centauri system as a triple-star system, Lalande 21185 was in fact the sixth-closest star, with Barnard's star and Wolf 359 being the two closer to the Sun, after the Alpha Centauri triple-star system. However, Gatewood made the mistake of continuing to take data on Lalande 21185 so that by 1998, his evidence for the 6-year-orbital-period planet had disappeared. Shortly thereafter, in 2002, George and I joined forces to start the CAPSCam astrometric planet search effort, with plans for a new astrometric camera on the 2.5-m du Pont telescope in Chile. By then, George's earlier claim for two planets around Lalande 21185 was no longer a topic for polite conversation between friends.

So what had Paul found around Lalande 21185? Had George really found something with astrometric observations taken with the ancient 30-inch refracting telescope at Allegheny? Well, no. Paul's Doppler observations had found a hot super-Earth orbiting Lalande 21185, a planet with a mass at least 3.8 times that of Earth and a 9-day orbital period. Such an exoplanet could not possibly have been detected by Gatewood's astrometric search. Whether Lalande 21185 harbored the cold, long-period gas giants that George thought he had detected would remain to be decided by some future exoplanet search. Lalande 21185 lay too far north to be observed by our Southern Hemisphere CAPSCam search. In spite of being so close, as a red dwarf star with a mass only half that of the Sun, Lalande 21185 was too faint to be seen by eye, yet paradoxically, it was too bright for the CAPSCam camera, which had been optimized for even fainter red dwarfs. Given the discovery of its hot super-Earth, hinting at what else might orbit the dwarf at greater distances, Lalande 21185 would be a prime target for direct-imaging searches by the next generation of ground- and space-based telescopes.

The Fourth Time Is the Trick: NASA Ames once again hosted the biannual Kepler and K2 Science Conference (KepSciConIV), held on June 19–23, 2017. In order to take advantage of the expected large crowd of exoplanet astronomers at KepSciConIV, we arranged to hold our summer ExoPAG meeting on the Sunday before KepSciConIV started, and invited the KepSciConIV folks to arrive a day early and attend our 16th ExoPAG meeting in nearby Mountain View, California, on June 18. K2 was finding dozens of new exoplanets, but the clock was ticking, as it was estimated that the Kepler spacecraft would run out of thruster fuel in the spring of 2018. That would mean the end of even K2.

ExoPAG meetings routinely include an update from NASA HQ with the view from deep inside the Beltway. We heard that things were not going too well for the APD in terms of the current plan for the FY 2017 budget, which was still in continuing resolution mode in spite of the current FY being nearly finished. The House and Senate conference committee report had given APD an overall cut of $47.4 million, an 11% reduction across the Division. Paul Hertz would have the distinct pleasure of deciding how to handle such a drastic decrease, should it become law. More specifically, the language stated that WFIRST would have to fit under the total mission life cycle cost cap of $3.5 billion proposed by the Senate appropriations committee back in April 2016. On the positive side, hopefully that figure would provide some breathing room for the WIETR exercise, which appeared to be thinking that WFIRST would have to come in under $3.2 billion. On the negative side, the language supported technology development for a future star shade mission without specifying if any funds should be allocated to this development. That also would be Hertz's problem to solve.

Rus Belikov gave the closeout presentation for his SAG 13 activity, once again making the point that his crowdsourced estimate of the frequency of Earth-like exoplanets, meaning both Earth-size and Earth-like temperatures, was about 60%, the same number he had been carrying along for the last year or so. Eta_Earth_SAG13 was equal to 0.6, and that was that. It was a key number for consideration by the two Large Mission STDTs that planned on searching for and characterizing habitable worlds: the HabEx and LUVOIR teams now had the best number currently available for use in estimating the exoEarth yields to be expected from the mission concepts being studied.

But what about the Kepler team's own estimate of Eta_Earth? The Level 1 requirements for the Kepler mission, the legally binding science drivers, did not require a specific evaluation of Eta_Earth but rather only the "occurrence rates" for exoplanets of different sizes and orbital periods. Although seemingly the same thing, in reality they are subtly different. The difference was that the occurrence rates were largely a summation of how many exoplanets of different sizes and orbital periods had actually been detected by Kepler and confirmed by one means or another, without getting into the weeds about how many Kepler might have missed for some reason. That is to say, the crowdsourcing estimates of Eta_Earth performed by Belikov's SAG 13, where assumptions about detection rates and extrapolations from what was certain to what was unknown, were considered too radical an exercise to undertake for the Kepler Mission. The Mission evidently did not want to stake its stellar reputation on publishing a value for Eta_Earth_Kepler that might later need to be revised or updated, in spite of the fact that Bill Borucki's original intent for the forerunner FRESIP mission was to derive Eta_Earth, period. This crucial number had been demanded by both the Astro 2001 and Astro 2010 surveys as being necessary to determine before NASA could proceed to develop a terrestrial planet imaging telescope. Perhaps Belikov's value of 0.6 would suffice for addressing this critical need.

The Kepler team held a press conference on the morning of June 19, 2017, the first day of KepSciConIV. The team had finished their processing of the final batch of Kepler mission data, DR-25, which included all 4 years of data. DR-25 had yielded 219 new exoplanet candidates, of which 10 were considered to be Earth-size and possibly in their star's HZ. This brought Kepler's total number of Earth-like planets up to a total of 50, out of the grand total of 4,034 planet candidates. I watched the press conference on NASA television in my ExoPAG meeting hotel, then checked out and headed for the airport. No mention had been made of what the latest discoveries might mean for estimating Eta_Earth.

L3 Becomes LISA: On June 20, 2017, ESA approved LISA as a major mission to be launched in 2034. LISA was planned to have three separate spacecraft, arranged in an equilateral triangle, 2.5 million km on a side. In comparison, the ground-based LIGO detectors that found the first clear evidence for gravitational waves in 2015 each consisted of two, 4-km arms stretching out in perpendicular directions. LISA and LIGO would operate on the same principle: laser interferometry would be used to measure the precise distance along each arm, searching for the transient changes that would signal the passage of a gravitational wave. In the case of LIGO, near-simultaneous detections in both Louisiana and Washington state would serve as the proof of detections; whereas in the case of LISA, the fact that there would be three arms would serve as the internal consistency check: all three arms should show an interferometric phase shift, that is, a miniscule change in separation between the two spacecraft constituting that arm. The three spacecraft would fly along together around the Sun, lagging behind the Earth by about 20 degrees, separated from each other by about 7 times the distance of the Moon from the Earth.

The ESA decision to proceed with LISA had been relatively easy to make after the stunning LIGO discoveries. In addition, the basic technology needed for LISA had been proven to work in orbit by the LISA Pathfinder mission, and proven to work a thousand times better than was needed by LISA. LISA Pathfinder would be rewarded for this excellent work by being shut down at the end of the month of June. Such is the fate of robotic spacecraft once their prime and extended missions are finished. Unlike Kepler, Spitzer, and LISA, with their stable, Earth-trailing orbits around the Sun, LISA Pathfinder was placed into orbit at the L1 Lagrange point, directly between the Earth and the Sun. Because the first Lagrange point is not a stable orbital configuration, LISA Pathfinder needed to burn fuel periodically to perform what is called "station-keeping," maintaining a halo orbit around L1. Once LISA Pathfinder was turned off, it would eventually leave L1, bound for an uncertain journey elsewhere in the Solar System. It might even intercept the Earth at some future date. In order to avoid such a calamity, the LISA Pathfinder team used the last few drops of propellant fuel to nudge LISA Pathfinder away from L1 and onto a relatively stable, long-term orbit around the Sun, where it would be "put to sleep" like a faithful family dog beset with serious health problems.

PRD Part Deux: Guillem Anglada-Escude and his Pale Red Dot team were on a roll. Following the historic success of their 2016 search for and confirmation of Prox Cen b, Guillem announced on June 22, 2017, that they would restart their intense Doppler campaign on Prox Cen, looking for any siblings to Prox Cen b. Once again, the HARPS spectrometer on the 3.6-m La Silla telescope would be the workhorse, with ESO promising about 90 nights of 3.6-m telescope time. In addition to returning to Prox Cen, the team would expand their target list by adding in two more nearby stars, Barnard's star and Ross 154, the second- and seventh-closest stars, lying at distances of just 6 and 10 light-years, respectively, not that much farther than Prox Cen's 4 light-years. The Breakthrough Starshot initiative folks would have to pay close attention to what Guillem might find at these other nearby stars. ESO was eagerly anticipating learning more about the planets that Pale Red Dot II might find once the E-ELT was ready to start operations in 2024.

June 22, 2017, also saw the release of the names of the independent review team for the WFIRST exercise mandated by the Midterm Assessment. The WIETR clock was now running, with their final report due to be delivered by roughly August 22, if the 8-week timeline specified in the WIETR terms of reference was to be obeyed. With any luck, the WFIRST team would be back to work in a few months, back to preparing for entering Phase B, instead of sweating the details of their presentations to the WIETR review team.

Congress Rides, Nay, Gallops to the Rescue: The House Appropriations Subcommittee chaired by Representative John Culberson held a markup session on June 29, 2017, for the FY 2018 budget plan. Culberson's panel approved a markup that removed many of the painful reductions to science agency funding that had been proposed in the President's FY 2018 budget request. NASA's SMD would see an increase of 2% over FY 2017—a slight but significant improvement compared to the 1% decrease proposed by President Trump—while the top-line, total NASA budget would rise to a record level of $19.9 billion. NASA's Earth Science Division, however, continued to be the whipping boy, with an 11% cut proposed by the House. When the House markup bill appeared online on July 13, 2017, Culberson's favored Europa Clipper and Lander missions were included and given target launch dates of no later than 2022 and 2024, respectively. Mars 2020 was still in good shape, along with a $12 million mark for the Mars 2020 helicopter technology demonstration mission.

WFIRST would receive the full amount requested in the President's budget, $126.6 million, but the House noted the cost growth concerns being addressed by the WIETR review underway, and directed NASA to brief the House Committee on the results. Culberson evidently had become a star shade enthusiast, as the language specified that $20 million of the WFIRST slice of the pie was to be spent on star shade technology development. Furthermore, the language stated explicitly that "The Committee expects WFIRST to accommodate the Star Shade technology demonstration mission." Culberson evidently realized that the CGI on WFIRST would not be able to discover or study Earth-like planets around even the closest stars, such as Prox Cen; with a 2.4-m primary mirror, WFIRST was just too small. But a large star shade could do the trick, when combined with WFIRST's modest-sized primary.

Why was Culberson so enamored with finding nearby exoEarths? The reason followed a few pages later in the House draft bill: "The Committee directs NASA to ensure that the United States is the first nation to launch an interstellar mission to the nearest Earth-like planet that shows evidence of extant life." There. He said it again, much as he had in the FY 2017 House Appropriations bill, but in even stronger terms this time. NASA had until July 20, 2069, to reach Earth 2.0, and needed to figure out a way to get a spacecraft there at 10% of the speed of light. The language instructed NASA to prepare an interstellar propulsion technology report, including a section describing NASA's plans to accelerate the work being done on a star shade for WFIRST. The star shade was intended to allow WFIRST to examine the atmospheres of "rocky Earth-like planets in the habitable zones of stable, long-lived stars out to a distance of 10 parsecs." It could not have been worded better if I had written it myself. The careful wording suggested that an astronomer had played a role in crafting the House language, but who?

While the House report once again cited the need for NASA to follow the directives of the NAS decadal surveys, it was clear that Culberson was already deciding that Astro 2020 should put a high priority on a star shade for WFIRST, whether they liked it or not.

In fact, the House bill went on to "encourage" NASA to work with the NAS on creating something new, a "permanent Decadal Survey for Exoplanet Exploration" for the next decade and beyond. Furthermore, NASA was "directed to follow the recommendations of this new Exoplanet Exploration Decadal Survey in developing America's long-term plans for systematic interstellar exploration missions to Earth-like planets harboring life in our galactic neighborhood." Ye gods. Suddenly the cost of a star shade for WFIRST, along with that for WFIRST itself, was a single drop in the bucket compared to the House's plan for American dominion over our block in the galactic neighborhood. This truly was a plan to Make America Great Again, the Roman empire of our little corner of the Milky Way galaxy.

Let the Breakthrough Starshot crowd try to top that one. With the full faith and credit of the United States Department of the Treasury potentially behind it, Culberson's bill was no idle dream, no Star Trekkie's hallucination. But would it survive the federal budget reconciliation process with the Senate? The Senate had yet to weigh in on the FY 2018 NASA budget, but given the otherwise dismal prospects for federal budgets, the House markup had to be considered as a sign for cautious, if only private, jubilation. The fate of WFIRST was still very much in play, with or without a star shade; and until WIETR had its say, NASA's flagship exoplanet mission, along with Culberson's dreams, were in jeopardy.

So What Does Paul Hertz Think?: The APAC held its summer meeting at NASA HQ July 19–20, 2017. The stunning language in the FY 2018 budget request from the House Appropriations Subcommittee was a lively topic of discussion. The language regarding the star shade in particular was reportedly driving the WFIRST program managers crazy. In order to retain their sanity, NASA managers were noting that the Senate had yet to weigh in for this world title bout, and it was expected that much of the House's language would not survive the slugfest of budget reconciliation. The idea of creating a permanent Exoplanet Exploration Decadal Survey also struck some raw nerves at NASA HQ, where program managers have a hard enough time meeting the periodic demands of Decadal Surveys and Midterm Assessments without having the real-time dosage of gratuitous advice that a permanent decadal survey might try to force down NASA's throat. Besides, the APAC fulfilled that role.

We heard the ritual update of the status of JWST, which was now entering the critical final phases of its construction. All the manufacturing of the various parts of JWST had been completed: the telescope and its instruments, the huge thermal sunshield, and the spacecraft bus. The remaining work consisted of integration and testing (I&T), and if everything fit together when assembled for the first time, and if everything worked when powered up for the first time, all would be fine and dandy. However, experienced space mission engineers knew well that this I&T phase was often where unforeseen problems suddenly appeared and bit you. Nevertheless, there were still 3.5 months of schedule reserve remaining to be used, if needed, slightly more than was usually expected to be held in reserve for GSFC flight projects but slightly less than the more conservative schedule reserves that had been decreed to be the rule for a project that was as important to NASA's reputation as JWST.

The summary stated that JWST was within the cost envelope demanded by the 2011 replan and was also on schedule for the October 2018 launch readiness date. However, a scary problem had arisen. Although JWST might well meet the replan's launch readiness date, there was the distinct concern that the Ariane 5 launch pad at ESA's Guiana Space Center in Kourou, French Guiana, might be occupied in October 2018 by a joint ESA-Japanese mission to Mercury. The BepiColumbo Mission, named after the Italian celestial mechanic who pioneered the gravity slingshot maneuver that allowed NASA's Mariner 10 spacecraft to fly by Mercury in 1974 and again in 1975, had been delayed in its development and construction and was scheduled to launch on an Ariane 5 from Guiana in October 2018. Unfortunately, the Guiana Space Center has a single launch site, Ensemble de Lancement Ariane 3 (ELA-3), capable of supporting an Ariane 5 launch. Given the complicated interplanetary route that BepiColumbo would need to take to reach Mercury, including one Earth and two Venus flybys, BepiColumbo had a restricted launch window. JWST was headed out to the second Lagrange point, L2, on the other side of the Earth from the Sun, and had no such restriction. BepiColumbo would occupy the launch pad for close to 6 months, as it was effectively three spacecraft mounted on top of each other, a transport module and two separate orbiters, and it would take some time to assemble and test the entire stack. JWST would need about 3 months to be mated to its Ariane 5 rockets and tested on the Guiana launch pad.

Depending on when BepiColumbo was delivered to Guiana and began its final stack assembly, there was a good chance that JWST's launch would be forced, by matters not under its or NASA's control, to delay its launch until the spring of 2019. Oops. Normally astronomers would be perfectly happy to wait a few more months for the launch of a flagship telescope as powerful and revolutionary as Webb was likely to be; but in this case, the real concern was the ongoing, close Congressional scrutiny. The FY 2018 House Appropriations Subcommittee bill released a few weeks earlier included explicit language requiring NASA to provide Congress with quarterly updates on JWST's "technical status, budget, and schedule performance, including program integration and tests that must be completed prior to its October 2018 launch." The Congressional language did not refer to a October 2018 launch readiness date, but to the October 2018 launch, period. NASA would have to negotiate with ESA about exactly when JWST might be launched, but the BepiColumbo hurdle appeared to be insurmountable.

TESS, on the other hand, was not expected to have any such problems with the launch pad (SLC-40 at Cape Canaveral) needed for its scheduled March 20, 2018, launch on a SpaceX Falcon 9. Like JWST, all of TESS's parts were now completely fabricated and ready for I&T. However, the I&T work on the TESS instrument camera had uncovered an unexpected problem with the focus of the four cameras that would perform the photometric transit searches that Kepler had pioneered in space. TESS, like Kepler, was intentionally designed to have slightly defocused, image plane CCD detectors, so that the brighter stars would not saturate the pixels at the center of their images, resulting in an incorrect count of how many photons had hit the CCDs. When one is trying to measure the brightness of a star to one part in a hundred thousand or so, one does not want saturated images and lost photons. At the same time, as the idea is simply to count all the photons from the star, and not to take a sharp picture, revealing any planets that might be in orbit, there is no need for a perfectly focused image; spreading the star's photons out onto a 5 × 5 grid of pixels would allow all the photons to be registered in the detector. But the problem that arose during testing was that once the camera was cooled down to the operating temperature in space of about 75° below zero Celsius (i.e., about 198 Kelvin), the focus changed across the image plane, with the central pixels becoming more in focus and the outer pixels becoming more poorly focused. The problem was even worse when the temperature was dropped another 10° Celsius.

Oops again. It was thought that the problem was due to partial crystallization of the rubber-like material used to bond the lenses to the rest of the camera. Rather than stopping and taking the flight camera apart in order to rebuild it, which would entail a major cost and schedule overrun for a class of low-cost missions (MIDEX) where termination would be a more likely decision, the TESS team and NASA HQ decided to proceed with the flight camera "as is," sort of like buying a house in "as is" condition: you bought it, and now it is yours, regardless of any flaws you might find afterward. The team would use the flight spare camera to continue testing and to figure out how best to mitigate this problem once TESS was in space. The decision to fly with a flawed camera was supported by the team's estimate that even with this problem, their expected yield of exoEarths and super-Earths would only drop by about 10% to 20%. The Level 1 requirements for TESS had been specified for detecting and measuring the masses of 50 small planets around nearby, bright stars. The team expected to discover several hundred new worlds, so even a 20% drop in sensitivity would not come close to forcing TESS to fail to meet its Level 1 requirements. In all likelihood, TESS would produce more transiting exoEarth candidates than the JWST telescope assignment committee would care to invest precious JWST time observing, searching for biosignatures.

We also heard from the co-chairs of the two Large Mission STDTs concerned with direct imaging of the exoEarths that TESS would find. Debra Fischer of Yale University noted that her LUVOIR team was considering two designs, one with a 9-m-diameter primary and the other with a 15-m primary. In either case, these mirrors would be too large to launch using any existing or planned launch vehicle; and as a result, they would be composed of folded, segmented mirrors, following along the path already laid down by the 6.5-m-diameter JWST. Scott Gaudi of the Ohio State University, and APAC chair, told us about the HabEx team's consideration of a 4-m-diameter monolithic primary mirror design, the largest size that could fit comfortably in existing rocket launch fairings. The exoEarth capabilities of this 4-m primary would be amplified immensely by flying a 70-m star shade in tandem with it. The HabEx folks were also considering a 6.5-m segmented primary, basically an ultraviolet and visible light copy of JWST. Both of the LUVOIR and HabEx designs were intended to cover the spectrum from ultraviolet to near-infrared light, allowing all four concepts to search for the biosignature molecules of oxygen, ozone, water, carbon monoxide, carbon dioxide, and methane.

News Flash: Brazil E-ELT Participation in Jeopardy: The July 21, 2017, issue of *Science* featured a breaking news article stating that Brazil's plans for participating in telescopes like the E-ELT were in jeopardy, to put it mildly. After imposing a 44% drop in funding in the current year's science budget, the Brazilian government was posed to decrease science funding by a further 40% in 2018. Egad. Those two cuts alone implied a total drop to about one-third of what the budget had been. The E-ELT had been expecting Brazil to fund a major fraction of the cost of the 39-m Chilean monster telescope.

ESO was still trying to find new partners. It was announced in October that Ireland was willing to join ESO, but the proposed annual Irish contribution to ESO of about $2 million would not go far in paying for the E-ELT.

16

Say, Could You Help Me Out?

Reality must take precedence over public relations, for nature cannot be fooled.
—Richard P. Feynman, 1918–1988

Scott Gaudi was off to Warsaw, Poland, to attend a conference celebrating 25 years of exoplanet microlensing discoveries made possible by the OGLE (Optical Gravitational Lensing Experiment) surveys running on the Warsaw University 1.3-m telescope at the Las Campanas Observatory. As a result, he asked me to take his place as chair of the APAC and to represent the APD at the NAC's Science Committee meeting being held at the same time as the Warsaw conference. After the APAC meeting ended on July 20, 2017, Scott, Paul Hertz, and I assembled the PowerPoint presentation that I would show at the Science Committee meeting the following week, on July 24, 2017, near NASA's Langley Research Center in Tidewater Virginia.

The Science Committee meeting was sparsely attended compared to similar meetings I had attended when I had been chair of the APS. When the Science Committee met at NASA HQ, the chairs in the back of the room would all be filled, and often a few NASA folks would be standing and listening as well. The only folks at the Langley meeting were the Science Committee members, the support staff, and a few invited speakers, even though this meeting, like all FACA meetings, was open to the public. Still, the meeting was available both by telephone dial-in and Internet conference services, so we made sure to speak into the microphones, unaware of just who the audience might be outside the room.

As we had agreed, I presented a quick summary of both the state of the APD and of the issues that concerned the APAC enough to be formulated as requests to Hertz in the letter reports that resulted from each APAC meeting. The Science Committee members listened to my presentations and asked numerous questions, including whether everything I was saying was intended for the public. I replied that everything was based on what had been shown and discussed at the APAC meeting—which, being a FACA meeting just like the Science Committee meeting, was open to the public. The Committee had picked up on the acute significance of both the JWST schedule danger posed by BepiColumbo and the camera focus problem facing TESS. We discussed those two issues and others and then broke for lunch.

The next day, during a joint session with the Human Exploration and Operations Committee, I spoke with a senior member of the human space flight community, who still resided in Texas near the Johnson Space Center and so was well versed in Texas space politics. We had been discussing privately the remarkable directions for NASA embodied in the FY 2018 bill produced by Texas Representative John Culberson's House Appropriations Subcommittee, including the language to figure out how to send a spacecraft to the closest habitable exoEarth by 2069. My colleague whispered to me that Culberson's House seat was thought to be in jeopardy for reasons other than his dream of having NASA accomplish not just interplanetary, but interstellar, space flights. Culberson was a strong supporter of President Trump, but his Texas district had voted for Hillary Clinton in 2016, giving hope to those who planned to run against him in 2018. Comparatively mundane political reasons might lead to the downfall of arguably the most visionary advocate of NASA space exploration we had seen since President John F. Kennedy sent us on our way to the Moon in 1961.

"Visionary" was one word that might describe Culberson's NASA dreams, but "imaginary" might be equally appropriate. When thinking about what the technology readiness might be for achieving the ability to accelerate a fully functional spacecraft to a speed of 10% of the speed of light, it was not clear if the usual natural numbers of 1, 2, 3, . . ., 9, used for TRL assessments, would be adequate: the complex number plane might be necessary, where the real number axis is supplemented by an axis consisting of the imaginary numbers (i.e., numbers that are products of the square root of negative 1). Perhaps the Breakthrough Starshot initiative crowd would offer a position to Representative Culberson if his competitors should unseat him in 2018, as the Starshot team was also fond of operating in the complex plane of interstellar space flight.

Don't Shoot the Messenger—Part II: Back in my office on July 27, 2017, I clicked on the NASA Watch web site for my daily dose of NASA gossip and was surprised to see my name in the day's top story. More folks had been listening to the Science Committee proceedings earlier that week than seemed to be the case at the time, given the meager numbers in attendance. A reporter from *Space News* had picked up on the problem with the TESS camera going out of focus once it was cooled to operating temperatures. Both a NASA spokesperson and I were quoted in the lead paragraph of the NASA Watch version of the *Space News* story, including my statement that the TESS team thought that the resulting exoplanet losses would be about 10%. An infuriated NASA Watch reader seized on my colloquial language as being imprecise and said, "Well if you are using my tax dollars you had better damn well know."

Well, as a matter of fact, I was not using your tax dollars, other than to pay for the mileage on my personal car, which I had driven down to Langley to participate in the Science Committee meeting, and for my other travel expenses. APAC members are considered Special Government Employees (SGE) on those days when they are working on NASA business and as a result have to fill out financial disclosure forms and jump through other annual bureaucratic hoops. However, we are not paid for this service to NASA. The same is true of my work as chair of the ExoPAG, with monthly telecons, daily emails, and two face-to-face meetings to organize and chair each year. Combined with the APAC meetings, and my agreeing to take Scott Gaudi's place at the Science Committee meeting, this meant that I was donating about a month of my time each year to serving NASA. You're welcome.

Perhaps the Internet troll, who did give a name, had thought I was a member of the TESS team, but a careful reading of the underlying *Space News* story should have disabused the troll of that notion: I was speaking as a member of the APAC, not as a TESS team member. Even had I been a team member, though, I could not have been more precise than I was, as this was a newly discovered problem, and the TESS team was still trying to figure out how best to ameliorate the situation. Even a 10% hit in the context of what TESS would achieve was not really a cause for major concern, as TESS would still uncover hundreds of exoEarths.

Space News presented a considerably more important story that same day, namely, the release of the Senate Appropriations Committee markup of the FY 2018 budget bill that included NASA. The Senate proposed a top number of $19.5 billion for NASA, less than the House's total of $19.9 billion. The Senate would raise spending on Earth Science and lower spending on Planetary Science, the exact opposite of the House proposal. There would be no missions to Europa, neither the Clipper nor the Lander, if the Senate got its way. In fact, there was nothing planned for Planetary Science beyond the Mars 2020 rover. The Senate was considerably kinder to the APD, with $150 million marked for WFIRST compared to the House's mark of $126.6 million, of which $20 million was to be spent on star shade development for WFIRST, leaving $106.6 million for WFIRST itself.

JWST's budget request was untouched, as expected, and the Senate language did not mention the long-planned October 2018 launch date, or refer to any launch date for JWST at all. Evidently the looming scheduling problem with BepiColumbo had been relayed to the Senate staffers, and the groundwork was being laid to avoid punishing JWST over a launch date that was not under NASA's control.

Compared to the detailed language in the House markup, the Senate bill was terse. There was no mention of the need to develop the technology to accelerate a spacecraft to 10% of the speed of light or to find the closest habitable worlds that would be investigated by no later than 2069 by this interstellar speedster. Combined with the Earth versus Planetary budget battle, it was clear that the House–Senate budget reconciliation committee would soon be duking it out somewhere on Capitol Hill. Which missions would be the casualties of the anticipated brouhaha?

Not So Fast: Several emails arrived on August 1, 2017, spelling delays in planned searches for life beyond Earth. First came a NASA HQ email noting that their intention to release details about how to propose for building instruments for the Europa Lander mission would not be realized in August, contrary to plan. Worse yet, a new time frame for the release had not been established. Hummmhhh. Evidently NASA HQ was going to wait to see which missions emerged unscathed, or at least were still breathing, from the House–Senate budget reconciliation negotiations before making any further commitments to the Europa Lander mission.

That afternoon an upbeat email arrived inviting scientists to participate in a conference about the European PLATO mission that fall in England. PLATO's main goal was unchanged: using transit photometry, PLATO would search for habitable rocky planets around sun-like stars. However, there was an unstated change: the launch date was given as 2026, compared to the planned launch date of 2024 when PLATO was anointed as ESA's M3 mission in 2014. In the intervening 3 years, the launch date had slipped by 2 years, getting perilously close to the classic danger of a space mission whose launch date was slipping by 1 year each year. Not the same as Zeno's paradox, but equally frustrating.

Almost in passing, *Space News* posted a story that day noting the schedule problems with the launches of JWST and BepiColumbo, once again quoting me as the source for the startling news based on my Science Committee presentation. This latest story prompted a goodly number of online comments, snarky or otherwise, but at least no one seemed to be shooting in my direction as the stand-in messenger with the bad news.

Who Else Wants to Serve Their Nation (for Travel Expenses, Only)?: On August 4, 2017, I submitted several names to the NAS web site seeking nominations of knowledgeable scientists to serve on the ad hoc NAS Exoplanet and Astrobiology Strategy Committees called for in the 2017 NASA Transition Authorization Act. Unlike the Act, the formal NAS Statement of Task specified up front that the study reports would be inputs to the Astro 2020 Decadal Survey. No surprise there. But things were already running a bit behind schedule, as NASA's notional timeline called for the new committees to begin work by June 2017 in order to meet Congress's September 2018 deadline. It was now 2 months later than June, and evidently the committees had not yet been constituted. The fact that as chair of the NASA ExoPAG I had only been asked on August 1 by the NAS to submit names by no later than August 7, 6 days later, suggested that someone had decided it was time to get moving, now. I nominated a few likely suspects, then heard no more. I intended to keep low on this one.

No, You Are Not—Yes, We Are: Besides blocking the construction of the TMT on Mauna Kea, native Hawaiians had been actively protesting the construction of another major new telescope on a sacred summit, this time on Maui's Haleakala dormant volcano, just across a 48-km-wide strait from the Big Island. Haleakala hosts a number of telescopes, though not so many or as large as those on Mauna Kea; and the Haleakala telescopes are concentrated in a small area next to the national park that dominates the summit. The plan was to deliver the primary mirror for a new 4-m solar telescope on August 2, 2017, with the road to the summit being blocked on the night of August 1 as a precaution against the expected protests. Six people were arrested the next day for trying to block the delivery of the primary mirror, which was delivered intact and on schedule. The solar telescope was scheduled to begin operations in 2020.

The TMT project could only look on and hope that their request for a new building permit on Mauna Kea might be approved, following a recommendation in late July by a retired judge to the Hawaii BLNR that the permit be granted. TMT's opponents vowed to continue their legal challenges to the Mauna Kea location, in spite of the loss on Haleakala. So far, the score appeared to be tied, 1 to 1, though the final result of the Mauna Kea game was still being adjudicated.

Making Mars Great Again: The job opening for NASA's Planetary Protection Officer (PPO) closed on August 14, 2017. When first posted a month earlier, the job title led to a blizzard of Internet stories making fun of the position, as if NASA was seriously worried about being attacked by alien spacecraft. In reality, the PPO is responsible for ensuring that terrestrial microbes do not establish colonies elsewhere in the Solar System as a result of improper sterilization of robotic spacecraft, as well for establishing protocols for the safe return of extraterrestrial samples to the Earth for laboratory study.

The job ad was largely a result of an ongoing battle between JPL and the PPO over what the Mars Curiosity rover could sample on Mars, given that it had not met the highest standards of sterilization, and what the Mars 2020 rover would be able to do in the search for life, as it also was not planned to be as thoroughly baked as the Viking landers had been in the 1970s. Baking large spacecraft is expensive, along with designing them to be baked.

Someone decided to solve the problem by transferring the duties of the PPO out of SMD to a new home at NASA HQ that might prove more amenable to turning the Mars rovers loose. As it was, the Curiosity rover was prohibited from exploring the "special regions" where Mars orbiter images had noticed seasonal changes that might be linked to seepage of underground water. Unless a change was made in the plans for sterilizing Mars 2020, or a change was made in the PPO office, the Mars 2020 rover would be prohibited from sampling and caching rocks from these same special regions. JPL would not allow that to happen.

Thomas Zurbuchen told a NAS committee performing the Planet 2011 Midterm Assessment on August 28, 2017, that NASA was planning on launching a mission to Mars no earlier than 2026 that would be able to retrieve the samples cached by Mars 2020 and return them to the Earth for analysis about 3 years later. That set the marker for PSD's entrant in the race to find life beyond Earth: their horse would not cross the finish line until 2029. APD had plenty of time to get WFIRST in action by then.

Don't Shoot the Messenger—Part III: A prominent JWST advocate took it upon himself to chastise me in an email on August 19, 2017, for reporting the possible problems with JWST's planned launch date, public information that I had relayed to the NAC Science Committee during their meeting near Langley the previous month. I had been informed confidentially of several other instances where anonymous folks had complained about me in the context of JWST in my role as Chair of the ExoPAG, though the complaints had been checked out and found not to be substantiated. I was in fact as supportive as anyone could be that JWST should be launched successfully, and as soon as possible; WFIRST was waiting in line for the next flagship ride. I sent a polite reply thanking the astronomer for his comments, along with my best wishes, copying the NASA officials who had also received his email. I did not rise to take his bait, though in my mind the best wishes were in "air quotes." Keep in touch.

JWST was being tested in the huge thermal vacuum chamber at the Johnson Space Center (JSC) to the southeast of Houston in August 2017 when Hurricane Harvey parked itself over the Houston metropolitan area for close to a week. Houston received a record amount of rainfall, trapping residents in their homes and flooding the streets and interstates. JSC was shut down for over a week, with only emergency crews allowed on the Center grounds. I cautiously inquired of NASA HQ if this had meant any further schedule slip for JWST, as the remaining schedule reserve was dwindling, and I was reassured that everything was fine, just fine. I resisted making any further remarks about JWST and hurricanes.

Making NASA Great Again: Rumors were circulating in mid-August 2017 that Representative Jim Bridenstine would become the next NASA Administrator. NASA had been running in place with Acting Administrator Robert Lightfoot at the helm since Charlie Bolden stepped down in January 2017. Six months later, the new Administration found the time to start thinking about who should lead NASA, a decision that would likely be tied up with where NASA would be going next: back to the Moon or to cis-lunar space, pit stops on the way to Mars? These were all "TBD" in the absence of a new Administrator.

Lightfoot sent a formal email to NASA employees on September 1, 2017, alerting all to the nomination of Bridenstine as the 13th NASA Administrator. What would Bridenstine mean for the search for life beyond Earth? No one seemed to have a clue about that, though Bridenstine was thought to be a supporter of commercial space flight and in favor of a return to the Moon. Perhaps we would learn more during his Senate confirmation hearings, perhaps not.

Making Alpha Centauri Great Again: Jon Morse was back in the spotlight, announcing on September 6, 2017, that he and his colleagues, including the SETI Institute, had launched a crowdfunding effort to build a direct-imaging space telescope that would search for habitable exoplanets in the Alpha Centauri triple-star system. Their goal was to raise at least $175,000 in order to get started on designing the system architecture. Given the numerous previous NASA studies about the cost of such a mission, all in the range of $1 billion or considerably more, that meant that once that first $175,000 was raised, they would only need, let's see, oh, at least $999.825 million more in crowdfunding to build their telescope. Good luck with that.

17

The Pre-Decadal-Survey Decadal Survey

Webb's spacecraft and sunshield are larger and more complex than most spacecraft.

—Eric Smith, JWST Program Director,
NASA press release, September 28, 2017

In spite of my efforts to deflect attention from myself and toward more deserving scientists for service on the NAS Exoplanet and Astrobiology Strategy Committees that would prepare reports for input to Astro 2020, I was invited to be a candidate for the Astrobiology Strategy Committee on September 21, 2017. My first impulse was to say no, thanks, I was planning on washing my hair; but after some reflection, I thought this exercise might be interesting: the Astrobiology Strategy Committee would be wrestling not just with searching for life on HZ exoplanets, but on Mars, Europa, Enceladus, you name it. If nothing else, this service would provide grist for my next book, as the Committee would not begin its work until after *Universal Life* was due to be submitted to the publisher in late 2017. Not only that, there was the excellent food served to NAS committees at the DC, Irvine, and Woods Hole, Massachusetts, study centers. I could not turn down three free meals a day, plus snacks.

That evening I sent an email agreeing to be considered for the Committee, noting though that I was expected to remain as ExoPAG Chair until April 2018, and on the APAC until April 2019, in case those might be considered conflicts. As it turned out, being on the APAC was indeed considered to be a conflict, and the NAS officer said he would check to see if an exception could be made in my case. It looked like I might get off the hook in spite of doing the right thing by accepting the invitation; besides, I did not really want to gain any more weight.

Score, or so I thought at the time. A week later a NAS staffer asked me to fill out a Doodle poll in order to help the NAS schedule a date for the first meeting of the Astrobiology Strategy Committee. I had not received word if my conflict of interest could be ignored, but I filled out the Doodle poll anyway. Sigh. It looked like I would have to watch my caloric input after all, in anticipation of gaining a few pounds at the Academy's expense.

The next morning, September 29, I was a bit late for a DTM astronomy group seminar by one of the Breakthrough Listen team members because I had just received a "Google alert" that my name had appeared in an Internet news story. I was stunned to click on the link and learn that NASA HQ had made a surprising announcement the previous day: JWST's launch would be delayed by half a year. October 2018, the date engraved in the stone of the 2011 JWST replan agreement with Congress, had become March to June of 2019. SMD head Thomas Zurbuchen stated that the problem was not with the telescope, still in the thermal vacuum chamber at JSC, but with the spacecraft and its enormous sun shield, which was in the process of "integration and testing" with the prime contractor, Northrop Grumman, in Redondo Beach, California.

The joint NASA-ESA agreement about JWST required NASA to inform ESA 1 year ahead of time when a date needed to be scheduled for the Ariane 5 launch from French Guiana. The JWST replan date of October 2018 meant that if NASA was going to demand that launch date, it needed to let ESA know by October 2017, which was a few days off. A September 27 presentation to the NSF had listed the launch date of October 2017, but that was September 27. On September 28, the date became spring of 2019. No mention was made in the press release of the previous launch date's conflict with the BepiColumbo launch: Northrop Grumman had resolved that scheduling problem all on its own. BepiColumbo could not be cast as the villain in this final act; the JWST launch slip was caused internally, not externally. There was no immediate reaction from retired Senator Mikulski, though I thought I heard a faint scream from the direction of Baltimore.

Given that the 2019 hurricane season starts on June 1, 2019, I would be hoping for a JWST launch early in the spring 2019 window, rather than late. (Just in case.)

The most important line in the press release was the statement that the half-year delay would not result in an increase in the total cost of JWST: the existing budget reserves would "accommodate" the change in launch date.

The AAS web site posted a story written on September 29, 2017, about the JWST launch delay downplaying the integration and testing problem that was cited as the culprit in the NASA press release. The AAS story stated that the real problem was BepiColumbo, which needed to launch between October 5 and November 28, 2018, in order to play the planetary billiards game that would allow it to end up in orbit around Mercury. The AAS story mentioned that the I&T process at Northrop Grumman would benefit from the launch delay. The spin on this story was worthy of what had been happening in Washington since the arrival of the new Administration: different news sources told different stories. But which one was the fake news?

TMT Gets a Thumbs Up (Again): The surprises continued on September 29. Reading the *Washington Post* that evening, I found a small article stating that the Hawaiian BLNR had once again approved a building permit for the TMT on

Mauna Kea. There was no hint in the *Post* article of what the TMT protestors would do in response.

Only for the moment we are saying nothing, to paraphrase a line from a letter sent to a lawyer in J. P. Donleavy's *A Singular Man*. Donleavy had died a few weeks earlier, on September 11, 2017, and he would have understood the inconclusive, unending battle between the TMT project and the Hawaiian protestors, just as he had suffered with the courts for years over the literary rights for his first novel, *The Ginger Man*.

A few days later, it became clear that the protestors planned to appeal the granting of the TMT building permit to the Hawaiian Supreme Court. What a surprise. If the protestors could continue to stop construction from starting by April 2018, they would win; TMT would then be forced to shift their observatory site to La Palma. Gracias, muchas gracias.

Stockholm Is Lovely in the Fall: Kip Thorne would be booking a trip to Stockholm sooner, rather than later. He and two colleagues were announced as the winners of the 2017 Nobel Prize in Physics on October 3, 2017, along with their thousand-odd colleagues who built and operated the LIGO gravitational wave observatory. LIGO had by now detected a total of four gravitational waves, and that was enough for the Nobel Prize selection committee.

Two weeks later, the LIGO team announced a fifth detection, GW170817, this time accompanied by a burst of light, signaling that the event was caused by the catastrophic collision of two neutron stars rather than by the black holes implicated in the first four detections. Oddly enough, the optical detection was first achieved with the Swope telescope, a modest 1-m-(40 inch) telescope, named after Carnegie astronomer Henrietta Swope, which was the first to be installed at LCO in 1971.

The parallels between the new fields of gravitational waves and exoplanets continued. The first gravitational wave with an optical counterpart was detected 2 years after the 2015 detection of the first gravitational wave. This was the fifth gravitational wave detected, and the optical counterpart provided the first confirmation by a different technique. The first transiting hot Jupiter was discovered in 1999, 4 years after the first exoplanet was discovered in 1995. This was the 10th hot Jupiter to be discovered, and the first time that a separate detection method confirmed the Doppler detection of a planet in orbit around a star (HD 209458; see TCU). HD 209458 b was the first system where one could measure the star's light dimming as a result of the transiting planet, similar to the brightening and fading of the collisional debris of the two neutron stars observed in GW170817.

The body count on the NASA Exoplanet Science Institute web page on October 3, 2017, stood at 3,513 confirmed exoplanets, mostly from Kepler, with another 4,496 Kepler planet candidates awaiting confirmation. A total of 584 stars had been found with more than one transiting planet. While the Nobel Prize selection committee had yet to deem exoplanet discoveries to be Prizeworthy, there was some consolation in that Bill Borucki's Shaw Prize was worth much more than what the three recipients of the 2017 physics Prize would each receive, so there.

The Kepler Mission was closed out successfully by October 5, 2017, meeting all of their Level 1 requirements. The team decided not to publish a formal Kepler Mission value for Eta_Earth, and NASA HQ accepted their decision. There would be no Eta_Earth_Kepler, but we had Eta_Earth_SAG13, equal to 0.6, which would have to take its place.

The National Space Council held its inaugural meeting on October 5, 2017, chaired by Vice President Mike Pence, at the NASM's Udvar-Hazy Center at Dulles Airport. Pence announced that NASA was being directed to prepare plans for human missions to the Moon, followed by human missions to Mars. NASA still had no Administrator, but Acting Administrator Lightfoot noted that the Deep Space Gateway, a cis-lunar outpost, could serve as a gateway to both the Moon and Mars. There was little mention of science during the proceedings: the motivation was presented as human exploration, period. A number of astronomers, including John Grunsfeld, had already begun to consider plans for how the Gateway might be useful for assembling space telescopes so large that they could only be assembled in space. If so, ATLAST might be on the Astro 2020 drawing board, or something even grander.

To WFIRST, or Not To WFIRST, That Is the Question: The APAC met October 18–19, 2017, where the Big News was Old News: the earliest possible launch date for JWST would be March 31, 2019. The JWST presentation laid the blame for the launch delay solely on the need for more time to perform the I&T of the spacecraft and sunshield deployment at Northrop Grumman—the resulting solution of the scheduling problem with BepiColumbo was a benefit of the launch delay, not the primary driver. We learned that given the launch delay, and the need for 6 or so months of commissioning once JWST reached its orbit at L2, the first science was expected from JWST to appear in early 2020, barely in time for the final deliberations of the Astro 2020 Survey Committee, slated to begin in May 2020.

Paul Hertz presented the APD budget plan (see Figure 17.1) and informed us that there was a serious problem with the sophisticated infrared detectors that the United States would provide for ESA's Euclid dark energy mission, a problem that was only discovered when the detectors were cooled to their operating temperatures. Addressing this problem would force a launch delay for Euclid of at least a year, to perhaps 2021 or later. The TESS team, which had also encountered a testing problem when their cameras were cooled to operating temperatures, had continued studying their focus problem and how best to minimize the effects. Their conclusion was the same as we had heard at the previous APAC meeting: there would be only a minor decrease in the number of exoEarths that TESS would discover. TESS was still good to go for launch on March 20, 2018, though a delay in certifying the SpaceX Falcon 9 for use by TESS would later push the launch date out to April 16, 2018.

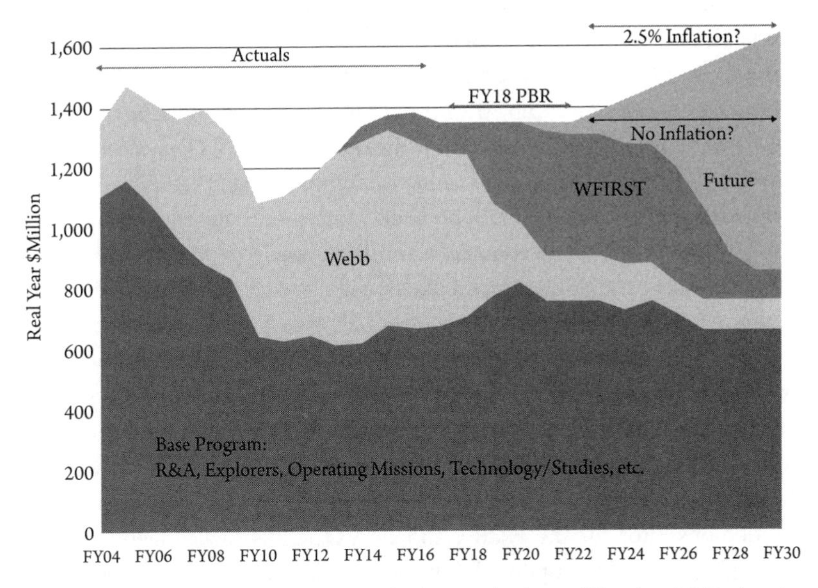

Figure 17.1 NASA Astrophysics Division budget outlook as of October 2017, showing the budget wedges for JWST, WFIRST, and future flagship missions (Courtesy of NASA).

Most important of all, once the public FACA APAC meeting ended on October 19, 2017, the APAC and the NAC Science Committee participated in a private telephone conference in order to be debriefed on the results of the WIETR exercise that had been underway for the last several months. The WIETR final report was about 2 months late, and we were anxious to learn the fate of WFIRST—was it ready to begin Phase B, or was it still in limbo? We learned that the WIETR team had found significant problems with the WFIRST program with respect to the anticipated costs, risks, and schedule: there was no way that WFIRST would be able to accomplish everything that was planned with the funds likely to be available. The WIETR team specifically called out the coronagraph as being of concern, as the CGI had become a front-seat driver for WFIRST rather than a back seat passenger. Ouch, again. Now what?

Thomas Zurbuchen said during the debrief that despite the seemingly dire status of WFIRST, "This is fixable." We had been sent a link moments before the WIETR debrief started pointing us to a newly released NASA web page containing Zurbuchen's reaction to the WIETR report. Zurbuchen accepted the WIETR report and directed GSFC to "conduct a top-to-bottom cost-benefit assessment to balance scope, complexity, and the available resources," and to do this before WFIRST could be considered for the mission design review that would allow the program to transition into Phase B. Zurbuchen stated that the cost had to be reduced to a total of $3.2 billion, considerably less than the WIETR team's estimate of $3.6 billion. In order to do this, he directed that reductions be made in the wide-field instrument and the CGI, but that the CGI "shall be retained."

Zurbuchen did not say "must be retained," the magic phrasing for NASA, but "shall" was not too bad. Dropping the option for star shade compatibility could save about $50 million, and "downscoping" (while retaining) the CGI could save another $50 million, so those two looked to be likely victims of the upcoming WFIRST replan. That still left another $300 million to cut, and the options for cutting the other instruments added up to only about $100 million. Would that be enough?

WFIRST was not the only project where costs had risen. Rumors began to circulate that while the GMT project was still trying to find more donors, the total costs of the competition had increased substantially, with the TMT now estimated to cost upward of $2 billion, and the E-ELT a cool $3 billion. Their costs appeared to be rising faster than new donors and partners were being found, a cycle that does not converge to an acceptable solution.

The plan for retaining the WFIRST CGI was to emphasize its role as technology demonstrator for the HabEx and LUVOIR mission concepts, to get the coronagraphic technology up to TRL 9, so that Astro 2020 would have no reason not to favor an exoEarth imaging mission on the basis of technological readiness, as Astro 2010 had ruled. The previous science requirements for WFIRST exoplanet imaging would have to be dropped. The new plan relaxed the requirement to achieve star-to-exoplanet contrast ratios down to 100 million to 1, deemed as being necessary for technology demonstration rather than for science. The wording for the new technical requirements used "will" rather than "shall," much less "must," a further weakening of the requirement language.

The new WFIRST would enter Phase B by April 2018, with a launch date likely to slip to March 2026. If the replan could not reach the magic $3.2 billion number, Zurbuchen threatened to start from scratch on WFIRST. That might be curtains for the CGI, which would be bad news for the HabEx/LUVOIR exoplanet folks. The HabEx STDT was including a star shade rendezvous mission along with purely coronagraphic mission concepts; and if the WFIRST project had to drop the star shade compatibility option, that did not bode well for the technological readiness to be anticipated for any future flagship star shade mission. Astro 2020 would get the chance to decide the issue.

Breaking News: November 2017 was a busy month, beginning on the 1st with Senate confirmation hearings for Oklahoma Representative Jim Bridenstine, nominated to become the next NASA Administrator. Though some Senators raised objections centered on Bridenstine's since-recanted 2013 remarks on the House floor regarding climate change, it appeared likely that he would win final confirmation and bring a fresh new face to 300 E Street SW. Bridenstine was gung-ho for NASA to send humans to the Moon on their way to Mars, but what would he think about the search for life beyond the Solar System?

The next day, Lamar Smith announced his intention to leave Congress and step down as chair of the House Science Committee. Smith has been a strong supporter of NASA, with special interests in astrobiology and SETI. He stated that the discovery of an "Earth-like planet that has methane and oxygen in the atmosphere" would be "one of the most astounding discoveries in humankind's history." (Applause.)

The following day, November 3, 2017, the GMT project began melting the borosilicate glass for its fifth mirror in the mirror-casting oven under the University of Arizona football stadium in Tucson. The surface of the molten glass will attain a parabolic shape as a result of the rotation of the mold inside the casting oven, a shape that is close to the desired mirror surface figure, simplifying the final polishing process. A photograph of the fifth GMT mirror being configured was featured in the 2018 calendar of the AAS, while the TMT image in the same calendar was merely an artist's conception, showing the interior of the dome, with no hint of where the TMT might be located. GMT was moving ahead, while still actively seeking new partners, planning to begin operations with four of the existing primary mirrors in 2023, and with all seven in 2025.

The winning Early Release Science (ERS) proposals for JWST were announced on November 13, 2017, with two of the winners planning to use 25% of the ERS time allocation to study exoplanets. Natalie Batalha and her team would have the first crack at using JWST to see if they could find evidence for biomarkers in the atmospheres of the transiting exoEarths around TRAPPIST-1 and LHS 1140. Natalie and her team were now poised to grab the brass ring of the first inhabited exoplanet, even if its inhabitants consist solely of methanogenic bacteria.

That was not all for November 2017. On the 15th, Xavier Bonfils of the University of Grenoble Alpes announced that he and his team used HARPS to discover a possibly habitable exoEarth around the red dwarf Ross 128, a star only 11 light-years (3.4 parsecs) away. Ross 128 b has a mass at least 1.35 times that of the Earth and receives about 38% more star light than the Earth, putting it at the inner edge of the HZ for Ross 128. Prox Cen is almost three times closer to Earth than Ross 128, however; so the Breakthrough Starshot folks would still be planning on checking out Prox Cen b first.

The December 22 issue of *Science* announced that the 2017 Breakthrough of the Year was the multiwavelength observation of the merger of two neutron stars, seen first by the LIGO gravitational wave detectors and nearly simultaneously by a wide range of gamma ray to radio wave telescopes, the first time that light was observed from a gravitational wave source. The discovery of the TRAPPIST-1 exoplanet system was not even mentioned as a runner-up, though TRAPPIST-1 did come in second to the top-ranked neutron star merger in Space.com's list of the "greatest space science stories of 2017."

More Breaking News: The status of NASA's response to the WIETR report was updated during the AAS winter meeting in early January 2018, held in the

Gaylord National Harbor Hotel, next to a partially frozen Potomac river. The news was relatively good, considering the unspoken threat of a descope: the CGI was being recast as a technology demonstration instrument, with reduced costs as a result, and WFIRST would continue to be formulated to be "star-shade-ready." The HabEx and LUVOIR STDT teams could now let out a sigh of relief, though the final WFIRST redesign was still underway. I had agreed earlier to serve as the chair of the Independent Review Team that would be monitoring the progress of the WFIRST CGI though all four phases (A, B, C, D) of its birth and development process, so the news was a relief to me as well on a cold day.

The GMT folks held an evening open house at the AAS meeting, featuring an open bar, snacks, and short talks, a sure recipe for a successful event. The GMT mood was upbeat, even if they were still looking for someone with deep pockets. TMT held an open house as well. Their AAS booth had a handout that confidently claimed that the TMT "is under construction on Maunakea, Hawaii." Wait, what was that? Another TMT handout more accurately noted that both "Maunakea" and Roque de Los Muchachos were under consideration, without mentioning anything about the current status of the site decision process. Which would it be?

The week after the AAS meeting, the NAS Astrobiology Science Strategy committee held its first meeting at the Beckman Center in Irvine. As expected, the meals and snacks were excellent. I ate like a graduate student. We learned that the Exoplanet Science Strategy committee had two co-chairs, David Charbonneau and Scott Gaudi. David attended the Irvine meeting and presented an overview of exoplanet discoveries. Other presentations focused on Solar System abodes, such as Mars, Titan, Europa, and other possible ocean world satellites. Because both NAS committees share the question of detecting life on exoplanets, the plan is to work together in hopes of settling on a consensus recommendation for searching for life beyond the Solar System, and to do it in a couple of months. A joint committee session, held on March 7, 2018, helped to cement these ties.

Trump Weighs In: The Presidential Budget Proposal for FY 2019, released on February 12, 2018, included an astonishing death sentence: the top priority of Astro 2010, WFIRST, was to be terminated. Such a death knell would reverse decades of Administrative, Congressional, and scientific respect for the priorities established by the venerable NAS Decadal Survey process, about to start a new round with Astro 2020. The justification offered was that after spending $8.8 billion on JWST, the Administration did not see spending over $3 billion on WFIRST as a priority. Could it be that JWST would claim another victim? In addition, rumors were circulating that JWST was in trouble again. Egad.

The scientific community was aghast and horrified by the drastic proposal. Emails circulated wildly, but Paul Hertz was able to email a voice of reason, noting that while Trump's proposal would be a starting point for Congress, it would be up to Congress to decide the fate of WFIRST. Given the past strong Congressional support for Decadal Surveys in general, and for WFIRST in particular, the hope was that Congress would issue a full pardon for WFIRST's perceived sins. Hertz stated that the activities planned for WFIRST in FY 2018 should proceed as planned, and indeed, they would.

The WFIRST project replan had been finished, and beginning on February 27, 2018, the 4-day Mission Definition Review for WFIRST was held at GSFC. The presentations during that meeting made it clear that WFIRST had been thoroughly scrubbed following the WIETR report, and it was ready to be approved for Phase B.

WFIRST would be serviceable by robotic spacecraft, potentially extending its life beyond the 10 years allowed by the size of the station-keeping fuel tank. International participation was planned to help lower the total price tag to the magic $3.2 billion number. While the CGI would be used for 3 months in order to demonstrate the technology, if the CGI was as powerful as expected, it was understood that the CGI could be used for science too, not just TRL work. The star shade was now part of the baseline mission, but it was at the top of the list of possible mission descope actions. Considering Kepler's unpleasant experiences, I was pleased to learn that WFIRST would have not four but six reaction wheels, three more than necessary. That lesson had been learned.

WFIRST's primary mirror had been built in 2000, and the WFIRST launch was scheduled for September 2025, a mere 25 years later. First, though, WFIRST was ready to be promoted to Phase B.

Fool Me Once, Shame on Me: A February 28, 2018, report from the GAO warned that JWST was in danger of additional launch delays, which would force it to break its 2011 cost cap. Northrop Grumman was having troubles with the thrusters and the sun shield, which was found torn in places. Simply refolding the sun shield requires almost two months, and there was little schedule reserve left. NASA HQ was left to estimate that JWST might not launch until the early fall of 2019, instead of June 2019. On the morning of March 27, 2018, I learned on a non-FACA APAC phone call that it would be announced later in the day that in fact JWST's launch date would be delayed even more than feared, to May 2020, or possibly later. The cost implications of this additional year of delay were unknown, and seemed likely to impact the ramp-up of WFIRST funding.

Incredibly, NASA HQ did not have an Administrator: Bridenstine's nomination was still unconfirmed. On March 12, 2018, Acting Administrator Lightfoot announced his intention to step down on April 30, 2018, whether there was a replacement or not. In the absence of an Administrator to mind the shop during the Trump presidency, we would all have to rely on Congress to keep WFIRST afloat in spite of the fiscal waves still being generated by JWST.

This Just In: TESS launched April 18 . . . Bridenstine approved as NASA Administrator April 19 . . . JWST drops screws and washers when moved to new testing chamber . . . House CJS appropriations bill increases NASA's FY 2019 budget May 16 with $130 M for WFIRST and $20 M for star shade technology development . . . WFIRST approved for Phase B May 22 . . . film at eleven . . .

Finale

It seems to me that what can happen in the future is . . . that the experiments get harder and harder to make, more and more expensive . . . and it [scientific discovery] gets slower and slower.

—Richard P. Feynman, 1965,
The Character of Physical Law,
Cambridge, MA, p. 172

There was a long interval from the 1993 cancellation of the SSC until the LHC discovered the Higgs boson in 2012. While Feynman's insight might have been prescient in 1965 with regard to the future of subatomic physics, it certainly does not seem to apply to the search for life beyond our Solar System, which only got started in 1995 in a serious way and has been accelerating in its discoveries ever since. The next steps to be taken, though, are becoming increasingly expensive, a common refrain throughout this book.

The competitors that will engage in the next phase of the habitable exoplanet discovery race between ground-based and spaced-based telescopes are now well known. NASA launched TESS on April 18, 2018, and plans to launch JWST over three years later, sometime in 2021. Their combination should yield information about the atmospheres of a few super-Earths in orbit about nearby red dwarf stars. JWST will be studying the TRAPPIST-1 system as soon as it can, and LHS 1140 b as well, looking for signs of habitable planet atmospheres among these worlds. The JWST delay should allow TESS to add a few more to the list. JWST stands the best chance of winning the race to find evidence of a few biosignatures in the atmospheres of nearby habitable, transiting worlds.

WFIRST is slated for launch in September 2025 and should produce a new microlensing planet survey, as well as offer unprecedented direct-imaging discoveries and atmospheric studies, thanks to the coronagraph. There might even be a star shade in WFIRST's future, should the Astro 2020 Decadal Survey bestow its blessing. If so, then WFIRST will be capable of imaging and characterizing a number of nearby Earth-like worlds orbiting sun-like stars. JWST and WFIRST should both easily beat the arrival on Earth in 2030 of Mars samples collected and cached by the Mars 2020 rover. Paul Hertz is likely to win his bet with NASA's new chief scientist Jim Green that APD will be the first NASA Division to find evidence for life beyond Earth.

Astro 2020 is scheduled to begin its deliberations in May 2019, and to deliver its final report to NASA and NSF in December 2020. The best guess is that APD would have about $5 billion to spend on new missions in the decade of the 2020s, perhaps enough to fly a flagship or two. Astro 2020 will then decide what will follow WFIRST: HabEx, LUVOIR, or something else? A gravitational wave detection mission will also be a strong contender for Astro 2020 to consider, with or without ESA. The next Astrophysics flagship mission(s) is purely TBD by Astro 2020.

Astro 2020 will also be faced with the unsettled issue of NSF support for the GSMT projects: will GMT or TMT receive any federal support? The ground promises giant strides forward in the next decade: the GMT is to have first light in 2024, followed by the TMT and the E-ELT, assuming that the TMT can find a good home. All three of these giant telescopes promise to make major advances in the discovery of alien worlds, both by Doppler spectroscopy and eventually by adaptive optics imaging. Prox Cen b will be one of their first direct-imaging targets: will Prox Cen b turn out to show signs of being habitable? If so, will NASA or the Breakthrough Starshot initiatives start work on sending mini-spacecraft to the Alpha Centauri system?

We will learn if it is really true that if it can be done from the ground, it will be done from the ground first. There is only one way to find out, and that is to run the experiment that the world's astronomers will be running for the next several decades.

We now know that Earth-like planets are universal, and we expect that life will be just as universal, even if it is primarily microbial, as Earth life was for most of its history. Considering the wide variety of exoplanets found to date, far beyond the imaginations of the most fertile science fiction writers, we can only dream about the weird life forms that might inhabit those worlds and about how equally weird we would appear to them. Some day we will meet them, and they will meet us, or our descendants, even if it is only through the medium of electromagnetic waves exchanged with each other across the otherwise impassable stretches of interstellar space and time.

LIST OF ACRONYMS AND ABBREVIATIONS

A&A	*Astronomy & Astrophysics* (European journal)
AAS	American Astronomical Society
AAAS	American Association for the Advancement of Science
AFTA	Astrophysics Focused Telescope Assets (NRO telescopes)
AIM	Astrometric Interferometry Mission (NASA)
AJ	*Astronomical Journal* (AAS journal)
ALMA	Atacama Large Millimeter Array (Chile)
AO	adaptive optics
APD	Astrophysics Division (NASA)
ApJ	*Astrophysical Journal* (AAS journal)
APAC	Astrophysics Advisory Committee of the NASA Astrophysics Division
APS	Astrophysics Subcommittee of the NASA Advisory Council's Science Committee
Astro 2010	2010 NAS Decadal Survey for Astronomy and Astrophysics
ATA	Allen Telescope Array (Hat Creek, California)
ATK	Alliant Techsystems Inc.
ATLAST	Advanced Technology Large-Aperture Space Telescope
AURA	Association of Universities for Research in Astronomy
BLNR	Board of Land and Natural Resources (Hawaii)
Caltech	California Institute of Technology, Pasadena, California
CAPSCam	Carnegie Astrometric Planet Search Camera
CATE	Cost, Risk, and Technical Evaluation (NASA)
CCD	charge-coupled device
CfA	Center for Astrophysics (Harvard-Smithsonian, Cambridge, Massachusetts)
CGI	coronagraphic instrument (WFIRST, GMT)
CHEOPS	Characterizing Exoplanets Satellite (ESA)

CJS	Commerce, Justice, Science (U.S. Senate Appropriations Subcommittee)
COPAG	Cosmic Origins Program Analysis Group (NASA Astrophysics Division)
CoRoT	Convection, Rotation, Transits (French-led space telescope)
CTIO	Cerro Tololo Inter-American Observatory (Chile)
DPS	Division of Planetary Science (AAS)
DRM	Design Reference Mission
DSIAC	Decadal Survey Implementation Advisory Committee (NAS/NRC)
DTM	Department of Terrestrial Magnetism, Carnegie Institution (Washington, DC)
E3LT	Exoplanets in the Era of Extremely Large Telescopes (Pacific Grove, California)
E-ELT	European Extremely Large Telescope
EOS	Electromagnetic Observations from Space (Astro 2010 panel)
ERS	Early Release Science (JWST)
ESA	European Space Agency (HQ, Paris, France)
ESD	Earth Science Division (NASA)
ESO	European Southern Observatory (HQ, Garching, Germany)
ESPRESSO	Echelle Spectrograph for Rocky Exoplanet and Stable Spectroscopic Observations (ESO VLT)
Eta_Earth (η_E)	Eta Sub Earth, the frequency of Earth-like planets around solar-type stars
ExEP	Exoplanet Exploration Program (NASA Astrophysics Division)
ExO	Exoplanet Observatory (NASA)
Exo-S	Exoplanet Star Shade (NASA)
exoEarth	Earth-like planet orbiting another star
ExoPAG	Exoplanet Exploration Program Analysis Group (NASA Astrophysics Division)
exoplanet	planet orbiting a star other than the Sun, i.e., an extrasolar planet
FACA	Federal Advisory Committee Act
FEMA	Federal Emergency Management Agency
FRESIP	Frequency of Earth-Sized Inner Planets (first name of Kepler Mission)
FY	fiscal year (1 October to following 30 September for U.S. Federal Government)
GAO	Government Accountability Office
G-CLEF	GMT-CfA Large Earth Finder (GMT telescope)
GMT	Giant Magellan Telescope (Las Campanas, Chile)

GMTIFS	GMT Integral Field Spectrograph (GMT first light instrument)
GPI	Gemini Planet Imager (Gemini Telescope, Cerro Tololo, Chile)
GSFC	Goddard Space Flight Center (Greenbelt, Maryland)
GSMT	Giant Segmented Mirror Telescope (NSF)
HabEx	Habitable-Exoplanet Imaging Mission (NASA)
HARPS	High Accuracy Radial velocity Planet Searcher (La Silla, Chile)
HAT	Hungarian Automated Telescope
HCIT	High Contrast Imaging Testbed (JPL)
HDST	High-Definition Space Telescope (AURA)
HIRES	High Resolution Echelle Spectrometer (Keck I telescope, Mauna Kea, Hawaii)
HST	Hubble Space Telescope (NASA)
HQ	NASA Headquarters (Washington, DC)
HZ	habitable zone (orbits permitting liquid water on an exoplanet)
I&T	integration and testing
ICRP	Independent Comprehensive Review Panel (NASA JWST)
IXO	International X-ray Observatory
JDEM	Joint Dark Energy Mission (NASA, Department of Energy)
JEO	Jupiter Europa Orbiter (NASA)
JPL	Jet Propulsion Laboratory (Pasadena, California)
JSC	Johnson Space Center (Houston, Texas)
JWST	James Webb Space Telescope (NASA)
KDP	Key Decision Point (NASA mission planning)
KITP	Kavli Institute for Theoretical Physics (UC Santa Barbara)
KOI	Kepler Object of Interest (exoplanet candidate)
L2	Second Lagrange point (on the Earth–Sun line away from the Sun)
L3	ESA's third large space mission concept (LISA)
LHC	Large Hadron Collider (near Geneva, Switzerland)
LIGO	Laser Interferometer Gravitational Wave Observatory
LISA	Laser Interferometer Space Antenna (ESA/NASA)
LCO	Las Campanas Observatory, Chile (Carnegie Institution)
LFE	*Looking for Earths* (previous popular book)
LUVOIR	Large UltraViolet Optical InfraRed space telescope concept (NASA)
MagAO	Magellan adaptive optics (LCO)
MagAO-X	Magellan extreme adaptive optics (LCO)
MAST	Barbara A. Mikulski Archive for Space Telescopes (STScI)

MAX-C	Mars Astrobiology Explorer-Cacher (2011 NAS Planetary Decadal Survey)
METI	Messages to Extraterrestrial Intelligence
METIS	Mid-Infrared E-ELT Imager and Spectrograph (E-ELT)
MIDEX	Medium-Class Explorer (NASA)
MIT	Massachusetts Institute of Technology, Cambridge, Massachusetts
MKO	Mauna Kea Observatory (University of Hawaii)
MPF	Microlensing Planet Finder (NASA)
MSL	Mars Science Laboratory (NASA)
NAC	NASA Advisory Council
NAS	National Academy of Sciences
NASA	National Aeronautics and Space Administration (HQ, Washington, DC)
NASM	National Air and Space Museum
NGAS	Northrop Grumman Aerospace Systems
NGST	Next Generation Space Telescope (NASA)
NOAA	National Oceanic and Atmospheric Administration
NRC	National Research Council of the National Academy of Sciences
NRO	National Reconnaissance Office
NSF	National Science Foundation
NWNH	New Worlds, New Horizons in Astronomy and Astrophysics (Astro 2010 Decadal Survey)
OGLE	Optical Gravitational Lensing Experiment (University of Warsaw)
OMB	Office of Management and Budget
OAO	Orbiting Astronomical Observatory (NASA)
OSI	Orbiting Stellar Interferometer (JPL astrometric telescope concept)
OSTP	Office of Science and Technology Policy (White House)
PCS	Planetary Camera and Spectrograph (E-ELT)
PRD	Pale Red Dot (Proxima Centauri team effort)
parsec	distance equal to 3.26 light-years or 31 trillion kilometers or 19 trillion miles
PhysPAG	Physics of the Cosmos Program Analysis Group (NASA Astrophysics Division)
PI	Principal Investigator (NASA)
Planet 2011	2011 NAS Decadal Survey for Planetary Science
PLATO	PLAnetary Transits and Oscillation of stars (ESA)
PFS	Planet Finding Spectrograph (Las Campanas, Chile)
PPO	Planetary Protection Officer (NASA)

PSD	Planetary Science Division (NASA)
PSSWG	Planetary Systems Science Working Group (NASA)
P Street	Administrative Headquarters (Carnegie Institution for Science)
RFI	Request for Information (NASA)
SAG	Study Analysis Group (NASA Program Analysis Group subgroup)
SETI	Search for Extraterrestial Intelligence
SGE	Special Government Employee (NASA)
SIM	Space Interferometry Mission (NASA)
SIM-Lite	Space Interferometry Mission (descoped option)
SIM-PH	Space Interferometry Mission (Planet Hunter option)
SMD	Science Mission Directorate (NASA)
SMEX	Small Explorer (NASA Mission)
SOC	Science Organizing Committee
SpaceX	Space Exploration Technologies Corporation
SPECULOOS	Search for habitable Planets EClipsing ULtra-cOOl Stars (Paranal Observatory)
SPHERE	Spectro-Polarimetric High-contrast Exoplanet REsearch (VLT, Chile)
SSC	Superconducting Super Collider (NSF)
SST	Spitzer Space Telescope (NASA)
STDT	Science and Technology Definition Team (NASA)
STScI	Space Telescope Science Institute (Baltimore, Maryland)
TAC	Technology Assessment Committee (NASA)
TAT	Test Assessment Team (NASA JWST)
TBD	to be determined
TCU	*The Crowded Universe* (previous popular book)
TESS	Transiting Exoplanet Survey Satellite (NASA)
TIGER	Thermal-infrared Imager for the GMT providing Extreme contrast and Resolution (GMT instrument proposal)
TMT	Thirty Meter Telescope (Caltech, UC-led private consortium)
TOPS	Towards Other Planetary Systems (NASA)
TPF	Terrestrial Planet Finder (generic NASA concept)
TPF-C	Terrestrial Planet Finder—Coronagraph (NASA concept)
TPF-I	Terrestrial Planet Finder—Interferometer (NASA concept)
TPF-T	Terrestrial Planet Finder—Transit (effectively the Kepler Mission)
TRAPPIST	TRAnsiting Planets and Planetesimals Small Telescope
TRL	Technology Readiness Level (NASA)
TTV	transit timing variations

TWOMASS (2MASS)	2 Micron All-Sky Survey (NSF, NASA)
UC	University of California
UCB	University of California, Berkeley
UCSB	University of California, Santa Barbara
UVES	Ultraviolet and Visible Echelle Spectrograph (VLT, Paranal Observatory)
VISIR	VLT Imager and Spectrometer for mid-Infrared (VLT, Paranal Observatory)
VLT	Very Large Telescopes (ESO's Paranal Observatory, Chile)
VSE	Vision for Space Exploration (President G. W. Bush's 2004 plan)
WFIRST	Wide-Field InfraRed Survey Telescope (NASA)
WIETR	WFIRST Independent External Technical/ Management/Cost Review (NASA)

INDEX